GETTING OUR WINGS

GETTING OUR WINGS

THE NAVY WAY

Bob Taylor

Cover design by Anthony Taylor

Author invites comments: roarta@hotmail.com
Website: www.bobwritesforyou.com

Library of Congress Control Number:		2015916297
ISBN:	Hardcover	978-1-5144-1388-3
	Softcover	978-1-5144-1390-6
	eBook	978-1-5144-1389-0

Print information available on the last page.

Rev. date: 10/21/2015

To order additional copies of this book, contact:
Xlibris
1-888-795-4274
www.Xlibris.com
Orders@Xlibris.com
725197

CONTENTS

DEDICATION

Without the hundreds of former and current Naval, Marine, and Coast Guard flight instructors, students, assistants, and civilian employees, as well as family members and friends who unselfishly shared their Naval Flight Training experiences and memories, this book could not have been written.

"Getting Our Wings" is dedicated to these men and women who so willingly shared stories of the good and the bad times, successes and failures, humor and sorrows, and sadness and hilarity of their training days.

ACKNOWLEDGEMENTS

Sincere appreciation and gratitude go to the PAO staff at NAS, Whiting Field, Florida. Hearing my proposal for this book were Jay Cope, PAO Chief, and Lori Aprilliano, PAO Officer at Whiting. Without hesitation, they set about locating documented history of the story of *NAS Pensacola.* They aided in my research for information and images of aircraft and training tools in use since 1911, the beginning of Naval Aviation.

Jay and Lori worked diligently in arranging quality interviews with training instructors, students, management, and staff. They scheduled demonstrations of the newest safety and teaching devices, foreign to those of us from days gone by. Their congenial and amiable assistance certainly were the keys to the completion of **"Getting Our Wings."**

PREFACE

Get ready to read one of the greatest stories that ever happened. It's about brave men and women climbing the ladder of courage to be ready to protect our nation in times of adversity. It's about the first 100 years of Naval Aviation training.

"**Getting Our Wings**" is an aviation book chock full of history, humor, nostalgia, sadness, and honor. Specifically, it is *THE* Naval Aviation Flight Training story, a non-fiction book about daring men and women spanning our first one hundred years as they learned to fly the Navy way. I would wager my last coin that many former flight instructors and students who read this book will think back to an earlier day and be certain that some of these memories happened to them during flight training days. "Hey, I believe that," they might say. "I think that's me. I'm in the book."

On this author's first day of Preflight in Pensacola in April of 1959, a Navy Commander herded us into a small darkened theater. We took our places in those soft seats – seats we were sure were usually reserved for flag officers and chiefs. We leaned back, anticipating a thrilling, patriotic, blood and guts, jingoistic message. The commander stood tall in his whites and threw out his chest, no doubt so we would have little difficulty beaming in on his gleaming Navy Wings of Gold. He began his welcoming speech to our class of 31 excited and apprehensive student flight officers — eighteen Navy, eleven Marine, and two Coast Guard. His words included a surprising bit of wisdom. "Statistically speaking," he said, "of every 10,000 babies born in the U.S. today, only one will satisfy the requirements to be accepted into Naval Aviation Flight Training." He paused briefly. Then with great emphasis, he leaned forward and stared into our eyes, "And that one is you."

His voiced changed as he warned us not to become haughty. He explained: (1) half the population of this country is female (that was 1959 — females were generally not candidates for Naval Aviation), (2) most young men are terrified to get more than a couple of feet above terra firma, (3) many young men would give anything to move to *Fly City* but do not meet the physical, mental, educational, or psychological demands, and (4) there are always those unmotivated individuals who just simply don't give a hoot.

"You're fortunate," he continued. "You were selected. Each of you is one in ten thousand. You are here in beautiful Pensacola." He paused for another meditative moment and half-raised his fisted right arm. "But now hear this. The rest is up to you. But not all of you will wear the Navy Wings of Gold. A couple years of heavy lifting is still ahead before you can cross that unyielding finish line."

His message was sobering as he pressed forward with a prediction that some of us in that room, having met the stiff requirements just to sit in these seats, for various reasons would not finish the demanding flight program. "Statistically, ten of you will leave Pensacola and not wear the coveted gold wings."

Spellbound and captured, we continued to listen as he ended his unusually short talk in dramatic fashion, "So whatever happens, gentlemen, hold your head up and throw your shoulders back. Never forget, whether or not you achieve your final goal, you are a member of the one in ten thousand."

"Wings" picks up Naval Flight training at the very beginning; back when it was self-taught. Self-taught? Pray tell, who could teach it? The reader learns that early pioneers knew very little about aeronautics. That word, by a hair's breadth, existed. Those aviation pioneers trudged, stumbled, tumbled, and died their ways forward.

Naval Aviation hatched in 1910, seven short years after the Wright Brothers lifted off into a freezing headwind on the early morning of 17 December 1903 from the wintry, cold beach at Kill Devil at Kitty Hawk, North Carolina. They were the first, most say. At any rate, they got the credit long ago. No doubt they realized they had done something special, but at that time could they have fully known just how massive it would be? Only we, who live a hundred years later, comprehend.

The book is primed with personal stories, data, and substance related to Naval Aviation Training. A million thanks for the support provided by the United States Navy Public Affairs Offices from New York to Pensacola to Whiting to Corpus Christi. But most of all, without the contributions of former and current flight students, flight instructors, flight support personnel, and families, the reader might never hear so many interesting, down to earth, and little known stories of Naval Aviation Training.

It has been my goal to put a noble face on Naval Aviation Training. I was once part of this grand and glorious Naval tradition that continues today. One of the proudest moments in my life was when my wife pinned those shiny new Naval Aviator Wings of Gold onto my United States Marine Corps uniform.

The reader will learn that the civilian populace was then, and remains today, important to maintaining a superior relationship between themselves and the Navy. Whether at Pensacola, Whiting, Saufley, Kingsville, Corpus, Beeville, or any of the many villages and cities harboring Naval flight training facilities, the Navy and the local citizens have enjoyed a history of friendship. They have worked tirelessly to maintain mutual respect and support. This is one of the primary reasons for success. As you read this book, you should agree that this camaraderie deserves every word of mention about the great relationship.

This book is not written in the conventional style of history. Better terms might be *heritage* or *legacy*. Compared with the word history, those terms sound more inspiring. Episodes are filled with allied documents such as legends, personal interviews, and stories of all varieties.

Stories drive this book. In selecting and editing these memories, stories and accounts, I endeavored to hold Naval Aviation to the high level it deserves, and to speak from the ground level to pass along memorable, nostalgic, humorous, real stuff stories, and yes, sad stories. I eased into play personal accounts and interviews from former instructors, students, training assistants, and families, oh so willing to help in this endeavor.

As I began the book, I felt sure that most people ever having been connected to Naval Flight Training could dredge up a story of humor, an entertaining story, a nostalgic event, or a story of dedication. I wasn't wrong. A few even opened their hearts and spoke of the heartbreak of rejection. When I asked, they shared, even at the risk of lowering their self-esteem in the eyes of the reader.

I cannot pass this moment without mentioning one of my favorite memories in the book; a story submitted by one of our country's heroes. He served in Tom Brokaw's "The Greatest Generation" and deserves his place of honor. **Dode** he is called. Specifically he is USMC Colonel George Dodenhoff, an over 90-year-old retired aviator living in Tampa with his perky wife, Priscilla. Dode, in an interview and later by written account, described one of the most beloved memories of his life. He dubbed it, "The Legend of the Avengers of Pearl Harbor." Certainly this is not a legend. It was real. It occurred one year after the bombing of Pearl Harbor. I know. Dode told me.

Dode's narrative demonstrates the patriotism of our country and certainly of our young men and women during those times. They were patriots who feared for the future of our country and its citizens. They laid themselves on the block to resolve the problem. Truly a spine-tingling story. As Dode talked, I listened, as did my heart. Tears probably welled in my eyes as will happen as to you and all who read it hence. Dode's full account appears in Chapter 7, "The Road to Pensacola." Please don't miss it.

As you prepare to read this book, escort it to your favorite easy chair. Sit comfortably as you look over the table of contents. The book is organized so that each time you pick it up, you may test your mood. Do you have a desire for heritage today? Then turn to the historical section, Chapters 1 through 6. Here are descriptions of how, why, and when we got there. Who knows? You will find yourself as you roam the pages.

Are you in the mood for: Humor? Drama? Reflection? Poignancy? Then hop over and begin with Chapter 7. You'll find first-hand humorous, entertaining, nostalgic, even heartrending true stories from the horses' mouths — those who lived them. You might have crossed paths with some of these former students, flight instructors, training assistants and families. Whatever your choice, enjoy them. Come back often.

CHAPTER 1

Our Heritage

Naval Aviation celebrated its first one hundred years in 2011. During that century, almost 300,000 Navy, Marine and Coast Guard Naval Aviators met the narrow prerequisites for selection, endured the demanding months of training, and earned their Navy Wings of Gold.

And such beautiful wings they are! If all of the world's top artists were to gather and struggle to reinvent them, they soon would suspend their efforts as being hopeless. Simply stated, the design of the Navy Wings of Gold is flawless. The beauty is matchless.

Orville and Wilbur

History credits bicycle shop owners Orville and Wilbur Wright, of Akron, Ohio, with the design, development, and the first successful flight of an aeroplane. At that moment, many others were trying. Some claimed to have succeeded. Others were still fighting to overtake them. The Wright Brothers' historical flight lifted off from a windy beach called Kill Devil near Kitty Hawk, North Carolina early one cold morning, on December 3, 1903.

The craft rose barely 20 feet off the ground and negotiated only 112 feet of horizontal air space before settling back to Mother Earth in a rather rough landing. Three days were required to repair the craft before its next flight. But the Wrights had achieved their goal. Kill Devil Beach remains a national shrine to commemorate this amazing achievement. What a beginning to manned flight; a flight so different from the ones we know today!

About this time, the Florida Naval Yard at Pensacola was becoming a relic. Our nation's warships seemly had no real purpose. To top it off, three years later in 1906, a brutal hurricane struck the area, followed by heavy rainfall and a damaging tidal wave, literally destroying the naval reservation and the harbor. Almost every vessel sank or beached. Streets were impassable. Impoverishment bashed the residents. With no work and no way to provide for their families, little hope remained. Today on the streets it is referred to as Pensacola's Tsunami.

The Navy announced plans to abandon the dying (or already dead) Pensacola Naval Yard and to move operations to Key West and Charleston, S.C. Matters could not have been worse for this West Florida community that had thrived on swelling employment numbers and a flourishing economy for several years. Pensacola, the city and the Naval Yard, fell into despair. Worst of all, the town and the Navy had always been, as many local citizens and Naval personnel would say, joined at the hip.

Meanwhile, as if planned by God, an experiment unknown to many, even those at the highest levels of the Navy, was underway at Annapolis, Maryland. In a large level and secluded field containing one shed and an ample supply of mechanic tools, the Navy Department was conducting secret experiments based upon Orville and Wilbur Wright's incredible success.

The Navy Department had signed a hush-hush contract for three flying machines to be built by the Glenn Curtiss Company and delivered to that little shed in Annapolis. Prior to these events, Curtiss had thrived on the manufacture of motorcycle engines. Now he was looking in a different direction in his fascination with flying vehicles. He wanted to be a part of what he hoped would be a new wave. Curtiss jumped on the bandwagon and would later be known as the "Father of Naval Aviation."

Even with such a frail status in the miniscule aviation business, Curtis looked ahead and saw a great future in winged craft. He proposed to high ranking doubters in the Navy that aviation actually was a bell ringer. Not everyone, a very few at first, listened but Curtis believed so strongly that he joined up with a man of like conviction, Eugene Ely, a civilian pilot, and they moved forward.

On 14 November 1910, Ely and his new partner traveled to Hampton Roads, Virginia and boarded the cruiser, USS Birmingham for a highly secret, enormously important, do or die breakthrough experiment. The two of them strolled down a newly constructed wooden platform lying over the deck of the vessel toward a 50-horsepower Curtiss-built airplane at rest. Ely climbed onto the seat, started the engine, and barked commands to his partner. Within minutes, Ely had accelerated down the platform and lifted off in successful flight. This must have set doubters afire with wonder.

Later, Secretary of the Navy George Meyer wrote a letter to Ely stating, "You are the first aviator in the world to have accomplished this feat. I congratulate you." Meyer continued his well done by telling the team that the flight was especially noteworthy. "That aircraft is old, underpowered, and aging. Because of the bad weather, the Birmingham could not get underway. Had she been at cruise speed, the supplemental wind assistance could have helped in the takeoff, thereby making this an even greater success."

Bustle within the Naval Aviation community picked up steam. More warm-body help was needed, so on 23 December 1910, Navy Lieutenant T. G. Ellyson was ordered to report to the

Glenn Curtiss Aviation Camp at North Island, California to work with Curtiss. Ellyson would become Naval Aviator No. 1.

A young Naval Officer, John H. Towers, had graduated from the United States Military Academy in 1906 and was immediately ordered to fill billets at sea. He began to hear of this new means of transportation invented by a couple of bicycle mechanics in Ohio. Wow! This is a vehicle that could actually attain flight and remain at altitude for long periods and cover a lot of territory. Amazing! That's what Towers wanted but he knew his captain would think he had lost his mind.

Meanwhile, another Naval officer, Lieutenant John Rodgers, received his selection to Naval Aviation. Being in the right place at the right time, Rodgers would become Naval Aviator No. 2 and was sent to the Wright Company at Dayton, Ohio to look over these new airborne marvels. In this short time, the Wright Company had progressed far beyond building bicycles.

Unfortunately, Towers still remained trapped at sea. He could only imagine how important such a device as an aeroplane could be if it were in consort with the surface forces. At present, when Naval vessels want to fire artillery shells at enemy forces on shore, visibility is often limited by ground fog or the curvature of the earth. Towers was convinced that Naval observers high in the air could benefit by the ability to see over the horizon. Their effectiveness could drastically increase. Having worked with manned balloons and airships, he saw the aeroplane as a possible solution to the dilemma of sometimes having to simply fire blindly. Besides, manned balloons, unable to maneuver effectively, were often shot out of the sky by the enemy.

However, the unknowing Towers wasn't sure of the legitimacy of aeroplanes and how long they might be a part of the Navy. He realized they might only be a flash in the pan. Why, he had never even seen an aeroplane! Still, the idea and what it could accomplish, if real, looked very favorable.

So the gutsy Towers put his reputation to risk. On 2 Nov 1910, he cautiously approached his captain, talked about nothings for a minute or two, cranked up a world of nerve, and hit him with his request. The captain, who had always expressed a dim view of aviation, in spite of holding a great deal of dependence and confidence in Towers as a surface sailor, flew into a rage. Why would his trusted officer dare to request reassignment into some crazy and unknown field known as aviation? The captain could not believe Towers would have such audacity to desert him.

Eventually Towers was accepted for Naval Aviation and transferred to Hammondsport on 11 June 1911. He would become Naval Aviator No. 3 and would report to Glenn Curtiss. He had hoped Curtiss would be his instructor, but to his dismay, T. G. Ellyson was assigned as his coach. Little did Towers know at that time that Ellyson was an equally proficient instructor.

Naval Aviation was in infancy. Little time and experience had been available to analyze proper training science. They were too busy flying! Their plane, a Curtiss A-1, was a one-seater. Ellyson could only instruct from the ground. Towers began his instruction by climbing aboard and familiarizing himself with the pedals but his heart longed to get that engine pointed skyward.

Soon Towers was told to climb into the cockpit for step two. Ellyson warned him that he could only taxi for the next few days. As a safety aid, Ellyson whittled a wedge of wood to squeeze under the aircraft's foot throttle. This wedge would be a *limiting device* to restrict the amount of throttle that could be applied. He told Towers that the plane could not take off, however hard he pushed the throttle.

Towers taxied around the field. After a while he excitedly jammed the foot throttle all the way down. In seconds the plane shot into the air, then back to the ground. Back into the air it went; then down and up. The baffled Towers had no idea how to control the airplane when finally it rammed headlong into a wooden fence. Towers was certain he had washed out on his first ever, and probably last ever, flight.

Finally the instructor aviators arrived at the answers to the problem. Ellyson was a hefty man but Towers weighed 30 pounds less. Ellyson's wedge should have been more restrictive. Less power was required for Towers to become airborne than Ellyson. That surplus power lifted Towers into the air. That might have been the day of the coinage of the age-old flight theorem that is well-drilled into the head of every student who ever goes to flight school: power controls altitude, attitude controls airspeed.

By September 1911, Towers, Rodgers and Ellyson had become more competent and were transferred to Annapolis to set up the new aviation camp. But after their planes arrived and flying began, Towers noticed something perplexing. Small holes began appearing in the aeroplanes. They were small, round holes about a quarter inch in diameter, the size of a student's writing pencil. But what were they?

It wasn't long before the mystery was solved. Out of nowhere the three new aviators arrived at the conclusion at the same time as they walked the perimeter of the field and heard gunshots from the rifle range. The flying field was directly in the line of fire of the range used by the cadets at Annapolis. They investigated further. Lo and behold, they were right. Midshipmen fired rifles on Tuesday and Wednesday evenings. On Wednesday and Thursday mornings, more holes were discovered. To further substantiate this, they found that the diameter of the holes were exactly the same as the bullets Midshipmen used in their rifles.

No one wanted to create any negative publicity about this. That might set back flying, so before the Navy would fly each day, aviators repaired the holes that had *mysteriously* appeared in the planes.

The existence of mysterious holes in the planes remained a secret. Only a few officers in high Naval circles were told — trusted ones. Naval Aviation wanted no floating riddles that might hinder this grand new invention. Thankfully, at the beginning of 1912, the aviation camp moved to North Island, just north of San Diego.

At North Island, other conditions plagued their flying. Unpredictable winds often flowed from the Pacific, sometimes causing very rough flying conditions. The aviators weren't familiar with these crazy whirlwinds. Their first analysis was that the turbulence was caused by moisture from recent rains boiling up from the ground into the warm California sun.

Towers, in attempting to fly one hot day, encountered some of the violent and unstable air. His plane was thrown thirty feet high then dropped thirty feet, repeatedly. After some extreme effort, he landed safely and Ellyson took off. In trying to counteract these aggravating conditions, he flew very low, just above the ground, in an attempt to avoid the turbulence. It didn't work. Ellyson was tossed from his seat, hit the ground and rolled over and over, knocking him unconscious. After several minutes he recovered and a doctor rushed up with a shot of whiskey. The flying machine fared better than Ellyson, but the Navy decided they would benefit with more flying and less repair expense if they moved back to Annapolis. In May they left, requesting a safe flying location, but certainly not one directly in the line of rifle fire.

A Momentous Flight Turns Heads and Goals

To many civilian as well as military personnel, flying somehow seemed more of a novelty. Naval Aviation was sort of touch and go. How could the top brass be convinced that flying was one of the most positive outlooks for the future? They argued that maybe Pensacola was not needed any longer. Why? Simply because the Navy was considering scrapping the base altogether. Then, an unforgettable and memorable experiment turned highly successful and extraordinarily valuable for the future of Naval Aviation and Pensacola. On the morning of 16 Feb 1913, Ellyson and Towers, both having been promoted to lieutenants, took off in a Curtiss pusher plane from Annapolis and flew south over Chesapeake Bay, a hefty 150 miles to Fort Monroe, Virginia.

A hundred and fifty miles! Heads turned and doubters began to lower their glances in disbelief. Everything was falling into place for the aviation enthusiasts. Within three years Pensacola was designated as an aeronautical station and the Naval Yard was reopened. This was a double whammy. The civilian community, remembering how splendid their relationship with the Navy had been before, embraced the decision. A state of symbiosis had returned.

Seat Belts, a Powerful Innovation

Seat belts for aircraft long preceded their use in automobiles. One windy day in June of 1913, Lieutenant Towers rode as a passenger in a Wright seaplane piloted by new pilot, Ensign William Billingsley. Suddenly, at 1,600 feet above Chesapeake Bay, the plane encountered severe turbulence. Billingsley was thrown from his seat and fell to his death into the bay. He became the first Navy pilot to make the supreme sacrifice.

Incredibly, Towers managed to catch and cling to a wing strut and rode the plummeting plane all the way down, miraculously surviving the crash. On that date, safety belts, previously thought to be little more than a cockpit nuisance, were ordered as standard and required equipment on all Navy planes.

Towers didn't back off after this close call. Neither did one of Tower's students, Patrick Bellinger, who was very vocal and straightforward about what it was like to fly these planes. He voiced his opinions using the most explicit adjectives: fear, exhilaration, weird, perilous, even death-defying. Bellinger was especially forceful about the times he was alone in the air, unassisted and solo. He admitted that he was nervous and excited at the same time and confessed that he actually knew very little about flying. He predicted, however, that the flight experiences of their day would, in the future, develop into the science of flight.

Bellinger spoke freely, spouting many of his convictions. He believed that none of the flyers knew much about flying. He said they made up the rules as they went along and that they knew that at any moment they were in the air, a simple ripple of the wind could toss them out of their seat to the earth below. He charged all pilots to pay constant heed and keep a sharp eye lookout constantly for good emergency landing sites. We of today remember that admonishment during our flying years.

Bellinger insisted that the motors were archaic and at any instant one might quit. Nevertheless, he actually had a basic trust in the mechanics. He had worked side by side with them often. He simply felt that flying was so new that not a single person could be trusted to get everything right.

In replying to Bellinger's tirades, Towers, his teacher, could not disagree. He confirmed that as of now, no science existed for flying. It was all trial and error. It would get better.

Readiness, 1917 Style

Naval Aviation was not ready for war in April 1917. Only one air station was fully operational. It had a complement of less than 50 aviators and about the same number of airworthy aircraft. Operational readiness was highly questionable. The thoughts were, War? We never considered that! Not in the air. However, the United States performed as it has always done when the cheese got binding. In those nineteen months between the call to arms to World War I and the armistice in 1918, performance was remarkable.

Training stations blossomed quickly in the U.S. and Europe. The Allies were actually coming to the ready. Soon, thousands of battle-ready fliers manned thousands of battle-ready aircraft to produce a formidable and combat-ready force. By the end of World War I, Navy and Marine Corps squadrons had formed the huge Northern Bombing Group, devastating the enemy in round-the-clock bombing raids.

During the course of this short war, aviation personnel realized that weather was one condition over which they had no control. The redeeming factor, however, was that the enemy was in the same boat. The Navy got smart. They grabbed hold of the idea that the U.S. must study and learn to adjust to adverse conditions quickly, and do it better than any future enemy. In November of 1919, the Navy opened the Naval Aerological School in Pensacola. This school still exists after several decades of metamorphism and name changes. Its status and eminence advanced by huge steps, being especially valuable for the needs of the new space program in 1957.

Lessons Learned but not Remembered Might Just As Well Not Be Learned in the First Place

Predictably, after this "War to End all Wars" ended, the world, and especially the United States, unfortunately relaxed, breathed and cut back on defense budgets. Why not? What else is new? The fighting was over and the Roaring Twenties were just beginning. It was to be a time for fun. Aviators were the darlings of the times. Air shows, including acts of parachute jumping and wing walking, flourished.

In the civilian world, criminal activities mushroomed, fed by the demand of citizens for the not-yet-legal booze. Speakeasies, also known as blind pigs or blind tigers, were the meeting places of choice for the high-rollers. Chicago, Washington, New York, Detroit, and other major cities had their gangs, led by hooligans such as Legs Diamond, the Genna Brothers, and Al Capone. Law enforcement was kept at bay by their bosses receiving special bequests such as huge sums of bribe money from under the table. When privileges didn't work, force did. Many an honest law officer (or his family) was found dead with no trace of responsibility.

Then in the late twenties, a massive financial crisis, the Great Depression, hit the country and ricocheted around the world. About the only well-off people were the mobs. This criminal element continued to thrive on their illegal clubs by paying off law enforcement and judges and hiring the best lawyers.

In 1932, Franklin Delano Roosevelt was elected president. He established programs such as the Works Progress Administration, hiring thousands of laid-off workers to build roads, bridges, and school gymnasiums.

Soon the world was drawn into the financial crisis. But Japan and Germany, the Axis Powers, began to rattle sabers again. Japan conquered and occupied countries in the Asian Pacific. They confiscated enemy wealth to pay for their ideals and conquests. Germany poured the same punishment onto countries in Europe.

Did anyone care? Obviously not! U.S. industrialists began selling war materials to the Axis Powers and coining money from all directions. The war materials were stockpiled in huge volumes and shipped helter skelter to countries that would turn them into munitions and pour them back onto friendly nations.

Readiness; 1941 Style

While the financial rendering down was underway, the U.S. turned its back on the intricate state of world affairs and the need for readying this country for battle. Then it happened. On the seventh day of December in 1941, Japan attacked us by hitting the U.S. squarely between the eyes with a crippling and sneaky blow to our fleet at Pearl Harbor, Hawaii. The Seventh Fleet, protectorate of our interests in the entire Pacific Ocean, was decimated, losing half its ships and aircraft and over 2500 military men.

In a foolhardy move for the Japanese and thankfully a positive one for the U.S., Japanese Admiral Yamamoto prematurely called off his attack planes. They retreated to their carriers and headed back toward their mainland. Even so, the U.S. ended in a real world of hurt. The U.S. leaders belatedly shouted into the air by sounding the call to arms. Many asked, "With what?" The major Seventh Fleet might that the U.S. possessed were the five carriers that just happened to be out of port on the morning of 7 December.

Six months later, a crippled fleet, including those five carriers, a few cruisers and destroyers, and many angry U.S. Sailors and Marines, set sail in a westerly direction to try to settle at least some modicum of a score. It could have been classified as a suicide mission. The Japanese were full strength and coming off a huge win, albeit with an ill-advised and premature conclusion.

In June of 1942, barely six months after Pearl Harbor, the two forces locked horns at Midway. The U.S. forces, knowing their days on earth might be numbered, fought heroically. The Japanese fought confidently as well; at least at first.

Was God Watching Our Six During Those First Six Months?

Maybe it was the great Kate Smith's nationally broadcast rendition on December 8, 1941, of Irving Berlin's "God Bless America." As she sang that musical symbol of patriotism, chills ran up and down the spine of every red-blooded American. As an eight-year-old boy living on a dairy farm in Georgia at that time, I didn't fully comprehend the whole concept of war. But I did understand the version that my parents imparted to me at the supper table that night and other nights to come. Two older males of my grandfather's family visited us and said they had quit their jobs and signed up to go to war.

I understood the concept even better a few weeks later on a rainy morning as my family drove one of those young men to the train depot. He boarded and waved to us through the windows. We waved back as my mother and grandmother cried as the train pulled away from the station. Angrily I waved my fists and yelled in a piercing demand to God, asking Him to let the war last long enough so I could "go kill some Japs." For that prayer, I received a misunderstood spanking.

Maybe our success at Midway just a few months after Pearl Harbor was Providential. It could have been just a plain miracle. Or it might have been sheer guts and determination. Almost every action the U.S. undertook worked with precision. Nearly everything that the Japanese attempted failed. We might say that they had a bad day at black rock. When the Battle of Midway ended, four of the five Japanese carriers, a cruiser, 248 Japanese aircraft, and over 3000 Japanese seamen rested at the bottom of the Coral Sea. For the last three years of the war, the Japanese, even though they fought bravely, remained on the defensive and never recovered. The great and mighty Japanese fleet was almost inert.

After the War, high-level United States and British military strategists met to review the War and especially the bombing of Pearl Harbor and the Battle of Midway. They remained puzzled that at Pearl Harbor the Japanese had the United States literally on the ropes. Why did Yamamoto draw back? Maybe only another day of their game plan might have struck the death knell to the U.S. They could have moved on toward Seattle, San Francisco and San Diego. It is obvious what only a few more bombing runs that day or another attack the next day could have accomplished. To describe it truthfully, we were toast. No reasonable answer was immediately present. They ended the seminar with no concrete answer. They only hoped the truth would someday come out.

Why Didn't the Japanese Make their Second Move?

Every history buff has his or her answer as to why the Japanese didn't take advantage of its Pearl Harbor gains. The following explanation has bounced around for quite a few years. It plays "what if" and supposes that if correct, the Japanese did their intelligence work well.

With all the Japanese living in the United States before Pearl Harbor some certainly must have been Japanese agents. These agents would have lived among the U.S. people for years, observing, collecting, and passing on information. They would know that hunting of all kinds was a popular sport. Hunters must obtain hunting licenses.

It would not have been difficult for foreign agents to determine the number of active hunting licenses. Wildlife Services publish this information every year. In easy strokes, Japanese Intelligence would know that for deer licenses alone, Pennsylvania had issued 750,000; Michigan, 700,000; West Virginia, 250,000; and Georgia, 500,000. The count for these states themselves alone add up to over two million – enough to make the point. It could be assumed that each license represented at least one American with at least one high-powered rifle or shotgun and that every owner would know how to use it.

Using the deer hunting license hypothesis, if the Japanese had landed on our shores, they might have found a swarm of millions of hunters eager to substitute the aiming point of their high-powered rifles from 8-point deer to groups of small easy-to-recognize foreign invaders.

Maybe that theory is a bit far-fetched. However, it does bear up our founding fathers' belief, along with the support of millions of Americans today, that the Second Amendment should be respected.

Another reason that Japan might not have moved toward our shores is logical and supported by historians. First assume that if territorial expansion had been a high priority objective, the Japanese would first want to occupy Hawaii and then move on toward our west coast.

However, occupying Hawaii or the United States was counter to Japan's goals at the time. The purpose of attacking Pearl Harbor was really defensive. By destroying Pearl Harbor, the U.S. Navy would not have been able to curtail Japan's occupation of the Philippines and many of the rich areas of the Pacific, such as Indonesia (then the Dutch East Indies). If Japan had occupied Hawaii and then moved on to the U.S. mainland, their entire homeland and their occupied territories would have been laid open to risk.

Yamamoto had to fight tooth and nail with the Japanese leaders just to keep the relatively small naval force assigned to his Pearl Harbor mission. The Japanese Army, always at odds with its Navy, fought against Yamamoto's plan. But when the Pearl Harbor mission succeeded so admirably, the Motherland's faith in Yamamoto amplified. But a mere six months later, June 1942, Yamamoto lost that faith when almost his entire fleet was slaughtered at Midway by the same Naval sea and air forces he had let slip through his fingers at Pearl Harbor.

Why are Pearl Harbor and Midway important to this book when its subject is United States Naval Aviation Training? Midway was, and remains today, the shining hour of our Navy and especially Naval Aviation. The 150 Naval and Marine pilots and air crewmen lost in the Battle of Midway trained as we trained. The same philosophies and flight maneuvers taught to them are taught today with certain exceptions for some aircraft. We must never lose that brotherly link with our forebearers who passed their legacy down to us. Those heroic men are very likely the reason that we live in a free country today.

Howard Hughes: A Man Unto Himself

One of our all time aviation greats was one of us; a noble and eccentric U.S. citizen, Howard Hughes. Actually more than eccentric, but extremely competent in aeronautics, Hughes gave U.S. aviation a boost during the nineteen forties and fifties.

As this book is written, flying machines (beginning with that Wright Brothers flight at Kitty Hawk in 1903) are into their 12th decade. In the 1940s, after over forty years in this new age of aeronautics, Howard Hughes had risen to the top. He was enticed by industrialist Henry Kaiser to make a "huge splatter". Why not? He was Howard Hughes. What else would he do? He began working diligently and in absolute secrecy. Word leaked out that a Hughes splash was imminent. The world waited. On 2 November 1947, the brazen and aggressive Hughes himself assumed command and piloted his secret and monstrous 200 ton HK-1 "Hercules," alias the "Spruce Goose," on its maiden flight over the waters off Long Beach, California.

The flight was intended only to be a taxi trial, but as aviators are prone to do when some strange urge comes upon them, Hughes and his engineering crew suddenly poured on the coal to the eight roaring radial engines, lifted off and gained an altitude of 33 feet and an airspeed of 80 mph. They held this monster jumbo of an airplane at 33 feet and 80 mph for a mile and then settled back onto the water in a perfect landing. Hughes had become the man who built and flew the largest plane ever to fly; an incredible accomplishment.

Some say the flight was accidental. The plane had not been tested and no one could vouch for how it would respond; or even if it would fly. But realizing how the spirit could grab Hughes, maybe this was just the point in time for which he looked. This flight was not only the maiden flight of the Spruce Goose, but her *only* flight. Many aviation experts of that time thought that Hughes might have simply wanted to silence the naysayers who popped up during the project. The project cost $22 million of which Hughes bore $18 million from his personal funds.

The Goose no longer launches into the sky. One flight is all she would need to permanently burn her name, and the Hughes name, into perpetuity. She rests at Evergreen International Aviation in McMinnville, Oregon.

Now we are one hundred years forward into Naval Aviation. This Marine has spent endless days, weeks, and months on Naval vessels. The part that amazes me is the part about a hundred years. The day I first walked up the plank of a Naval vessel, saluted the Officer of the Deck and turned to salute the National Ensign, Naval Aviation had barely celebrated fifty years. On that day, I realized that such a time span was a magnificent milestone. It is hard for me now to grasp that yet another fifty years have past since I first boarded a Naval vessel.

Has our honor deteriorated? No. True honor does not deteriorate. I think it assumed its place within each of us the first time we saluted Old Glory, whether as a kid reciting the Pledge of Allegiance with hand over heart or as a uniformed adult preparing to protect our country.

How about loyalty? Of course it has not declined. Real and honest to goodness loyalty does not weaken. Men and women who would leave their loved ones to spend months and years looking across the salty brine, or a foxhole, at an enemy who wants to annihilate us are prime examples of loyalty. Furthermore, those families waiting at home on the shores deserve equal acclamation.

Then maybe commitment? We would, as always, go to the ends of the earth for this country and its purposes. Courage? Never have the Naval Forces of the United States faltered. We remain determined. Give us a job. It will be done.

A New Dawning – Women in Naval Aviation

Naval archives tell us that the first licensed woman pilot in the United States was Harriet Quimby in 1911. History fails to mention that Katherine Wright, sister of the Wright brothers, had much to do with the first flight at Kitty Hawk as she worked with her brothers. Women flew airplanes before they could vote, but not in the U.S. military!

During World War I, Russian Princesses Eugenie Shakhovskaya and Sophie Alexandrovna Dolgorunaya were two of the first women to become military pilots in Europe. American women pilots had volunteered for combat but weren't taken seriously. We all remember the stories of the valiant WASP pilots. They flew every airplane made during WWII, including an experimental jet at 350 mph at 35,000 feet, (Ann Baumgartner in 1944). Strangely, these courageous ladies were not considered real military pilots until decades later.

Jacqueline Cochran broke the sound barrier in 1953. She set speed and altitude records and was a staunch lobbyist for assigning women pilots to the military. She could not break down the wall. Civilian women have even flown over the North Pole, circumnavigated the world, and broken the sound barrier but until the 1970s, the military continued to reject the idea of women pilots in military cockpits.

The Navy, not the Air Force, jumped into the ring first. In 1974 six women earned their wings and became the first female Naval aviators. The Army followed suit. They trained female helicopter pilots. Finally in 1976, the Air Force caught up and invited women into the pilot training program. But there is always a catch. By virtue of existing policies, their flying was limited to non-combat. Military women pilots would not be flying combat missions. At least, not at that time.

Although the military finally were training women pilots, the services still tried to work out the rules. Quotas had to be set. Budgets had to be met and women would be expending a portion. By law, women were excluded from combat and kept out of the cockpit of certain types of aircraft. Even though women aviators actually flew during the operations in Panama, Grenada and during Desert Storm, combat records did not indicate their being there. It was not until 1993 that women were given permission to openly fly combat missions.

The first woman pilot in the United States flew in 1911 but it took the military 65 years to recognize and train women as pilots and another seventeen years to permit them to invade the revered area of combat aircraft. The tide turned in 1997 when Air Force Colonel Eileen Collins led an all women flight, the Air Force Fly-Over Team at the dedication of the Women's Memorial. In 1999, Colonel Collins was selected as the first woman to command a space shuttle mission.

During Operation Desert Storm the first woman pilot was killed while flying in a combat zone. Air Force Major Marie T. Rossi died at age 32 on March 1, 1991, when her Chinook helicopter crashed near her base in northern Saudi Arabia. Her unit was one of the first American units to cross into enemy held territory as they flew fuel and ammunition to supply the 101st and 82nd Airborne Divisions. Major Rossi's epitaph in Arlington Cemetery reads "First Female Combat Commander to Fly into Battle."

Another American woman to fly combat missions in the 1990s was LtCol Martha McSally, who was ranked at the top of the Air Force female pilot list. She was one of the first to be trained by the Air Force as a woman fighter pilot. McSally was the first woman in military history to fly a combat sortie in a fighter aircraft. In 1995-96, she flew more than 100 combat hours in an A-10 Warthog attack aircraft over Iraq and served as a flight commander and a trainer of combat pilots.

In 1993, Secretary of Defense Les Aspin opened combat aviation to women. His decree included allowing enlisted females to fly as aircrew members and women pilots to fly combat missions. From this springboard, opportunities widened for women and crew-members as pilots. In the war against the Taliban and al-Qaida in Afghanistan and Iraq, women occupied aircrew positions as bomber pilots, navigators, tanker pilots, weapons officers, and the strategically important ground positions.

Top of Form

In June of 1995, President Bill Clinton spoke at Arlington National Cemetery on the occasion of the groundbreaking ceremony for the Women in Military Service for America Memorial, the first real national memorial honoring women who served in our nation's defense. Finally, equality had arrived for military women. Now their services may be remembered in a visible, touchable and beautiful memorial rather than simply by words in a history book.

Naval personnel will find it important that, in his speech, Mr. Clinton singled out Navy LCDR Barbara Allen Rainey the mother of two daughters. Rainey was the Navy's first female aviator, tragically the victim of a training crash. The President reminded the nation that LCDR Rainey's story strikes a chord in us that, even in peacetime, it is every day that danger faces those who wear the uniform. She now rests in eternal quiet and honor on those sacred grounds.

Thirteen years earlier, at Naval Air Station (NAS) Corpus Christi, Texas, LTJG Rainey was working her way up the ladder, completing task after task in her goal to receive her wings. On 22 February 1974, the bounds were broken when Rainey was designated and received her Wings of Gold in a graduation ceremony to become the first female Naval Aviator in history.

Those who have followed LCDR Rainey have reached heights that previously were unattainable. Women have flown combat missions off the decks of aircraft carriers. They have commanded combat squadrons. And certainly those who have launched into outer space as Space Shuttle astronauts are among the most heroic.

LCDR Rainey probably understood and accepted this heavy load for those to follow. She wanted women to be rated as equal to men in opportunities in all phases of Naval Aviation. It is certainly tragic that this woman upon whose shoulders these outstanding women pilots stood did not live to see her dreams of her gender to be fulfilled.

It was on that tragic day on 13 July 1982 that Rainey was a flight instructor in a T-34C Turbo Mentor based at NAS Whiting Field. She and her student, Ensign Donald Bruce Knowlton, were shooting touch and go landings at one of Whiting's outlying training fields. A second T-34C was also practicing touch and go landings. The two planes defied the age-old flying directive prohibiting two aircraft from occupying the same airspace simultaneously. The two planes collided, killing Rainey and Knowlton.

Today women in the United States Navy serve in almost every role on ships, in the air, and in command positions. Their contribution to our Navy is not a quirk. It is now a necessity. Barbara Allen Rainey prepared the way.

Brenda Bilger Robbins is another example of Women in Naval Aviation. Brenda was not an aviator, but worked in the Fleet during 1978-1983, training F-14/F-4 training squadron pilots at NAS Oceana, Virginia. Her first year on the job, she logged over 1000 intercepts and became the Navy's first female Air Intercept Controller Supervisor. Brenda notes that being the only female in the group produced no unusual consequences and that she was treated professionally and honorably.

Women pilots in Naval Aviation are recognized as having viable contributions to the security of our nation. Naval Aviator Lieutenant Commander Lena Buettner is a prime example. From February through August of 2004, her HSL-48 detachment deployed in support of Operations ENDURING FREEDOM and IRAQI FREEDOM aboard the USS LEYTE GULF (CG-55). The detachment included six pilots, three air crewmen and two lean and mean Seahawk war-machines as part of the Expeditionary Strike Group (ESG).

Serious threats surfaced indicating that insurgents aboard a "fishing boat" intended to blow two oil platforms. Lena and her crew proceeded in their assigned mission to protect these structures with their Seahawk armed with Hellfire missiles.

Their helo cover kept the enemy boat at bay until a Pakistani craft arrived and boarded the vessel.

As Lena and her crew set course for the Leyte Gulf, she received a change in mission to proceed to a Pakistani ship to Medevac a senior Pakistani military officer to the nearest medical facilities. Because of the critical nature of this mission, OPNAV restrictions were overridden which normally would prohibit her Seahawk from use on a Medevac mission in a combat configuration.

Buettner and her crew accomplished their assignment in time to save the life of the officer. The next day, the Fifth Fleet Admiral sent a formal naval message thanking Lena and her crew for performing the mission as ordered. The Fifth Fleet deemed that the high visibility of this mission was instrumental in strengthening relations between the United States and Pakistan during a time of political upheaval and uncertainty.

CHAPTER 2

NAS Pensacola

It all started in Pensacola. The Naval Air Station Pensacola, the Cradle of Naval Aviation, began with a few acres of land and a creative vision of explorers and settlers and progressed to its current status. It now occupies 5,804 acres and enjoys a significant historical background.

Its history dates back to the early 16th century when, in 1515, Don Tristan de Luna discovered the beautiful bay protected by highly fertile land to the north. Over a dozen years later, in 1528, the Spanish conquistador returned to the harbor with 2,000 settlers and soldiers and established the first white settlement in North America on Santa Rosa Island. Near the settlement lived a village of friendly Muskogee Indians who called their community Pensacola (loosely translated as "long-haired people"), the site now occupied by the Pensacola Naval Air Station. Since then the site has also been the location of Fort San Carlos de Barrancas, shortened to Fort Barrancas, a Spanish fort built in 1787. The flags of Spain, France, Great Britain, the Confederacy, and the United States have flown over the strategic port of Pensacola. The annual Fiesta of Five Flags continues to commemorate this history.

The 100th year of Naval Aviation is celebrated as this book is underway, but the Naval Base of Pensacola was actually hatched in 1821 when the United States purchased Florida from Spain. In 1825, President John Quincy Adams realized the strategic importance of Pensacola Bay as a support facility for naval squadrons operating in the Gulf of Mexico and the Caribbean and threw his support toward this end.

The Navy arrived in Pensacola that year and began preparations to build a Navy Yard at the precise location where Naval Air Station Pensacola stands today. By 1861, it had become one of the best-equipped naval establishments in the country. After being destroyed in 1862 during the Civil War it was gradually rebuilt but it was soon discovered that the Navy Yard could not hold its own against the developing steam industry. In 1911, the base was decommissioned and closed.

Later, with the practical development of aviation came the need for a permanent Naval Air Station. In 1914, the Pensacola Navy Yard was reopened and the name was officially changed from Pensacola Naval Yard to the Pensacola Naval Aeronautic Station, becoming the first air station in the Navy. On 20 January 1914, the USS Mississippi and a tender ship arrived in Pensacola from Annapolis, Maryland, and off-loaded the first supplies for the Naval Aeronautic Station. The original aviation unit to set up shop at Pensacola consisted of nine officers, less than two dozen enlisted men, and seven graceless flying boats, as well as hydro-aeroplanes commanded by Lieutenant Commander Henry C. Mustin. That was the start of the Navy's first air station and flying school.

Station logbook records began 10 February 1914, the date the air station was officially commissioned. The picture changed shortly with the entry of the United States into World War I. Within 2 years, Pensacola had trained over 1,000 aviators. During World War II, the number of pilots trained at NAS Pensacola reached its peak. Records indicate that 21,067 men completed the grueling and extensive training in 1944, the highest number of any year ever, and logged a combined total of almost 2 million hours.

Pensacola, the Strategic Hub

NAS Pensacola has always been known as the hub of naval aviation. Think about it. Your buddy says, "I went through

Pensacola." It's likely he doesn't mean he drove along I-10 from Crestview on the way to Mobile. Very likely it means he studied, sweated, struggled, toiled, and hustled for nearly two years in one of the toughest schools in the nation; one with the final prize of Navy Wings of Gold for every successful candidate.

NAS Pensacola manages countless educational and training activities and is propelled by more than 17,000 military and civilian personnel. The initial training of naval aviators begins here, but as this book is written, is shifting to NAS Whiting Field in Milton, Florida, thirty miles to the northeast. NAS Pensacola is comprised of over 70 Department of Defense and 25 other leaseholder commands primarily dedicated to the training of Navy, Marine Corps, and Coast Guard personnel in Naval Aviation.

Some of the major activities on board the air station include Chief of Naval Education and Training (CNET), Training Air Wing Six (TraWingSix), Naval Aviation Schools Command, Naval Aerospace and Operational Medical Institute (NAMI), and Naval Aerospace Medical Research Laboratory (NAMRL).

NAS Pensacola is currently the home of the Blue Angels (known officially as the U.S. Naval Flight Demonstration Squadron) and the National Museum of Naval Aviation. The Museum is free to all and is open year round every day except Christmas and New Year. Hundreds of aircraft dating from the beginning days of Naval Aviation may be touched, patted, admired, and photographed.

The headquarters of the Naval Education and Training Command, one of the largest Navy shore commands, is located at NAS Pensacola. Chief, Naval Education and Training Command (CNET) himself, is a vice admiral reporting directly to the Chief of Navy Operations. CNET oversees the training and education programs and activities extending from coast to coast and to ships at sea (as the famous World War II news commentator Walter Winchell might report). From the basics of recruit training extending through highly technical skills, instruction is included and covers over 4,000 courses. The mission of CNET changes daily as needs change to cover all possible teaching skills in maintaining a modern technical Navy.

The Wing Commander is overall commander of NAS Pensacola and NAS Whiting Fields, aviation training squadrons VT-4, VT-10, VT-86, and the 2nd German Air Force Squadron. Supported by the two air stations, the mission of Training Wing Six is to plan, supervise, support, and conduct quality flight training of Student Naval Aviators, Student Naval Flight Officers, Undergraduate Navigators, and International Military to satisfy service requirements. Among other responsibilities, TraWingSix also provides liaison between local operational units and the Chief of Naval Air Training, coordinates training airspace within the Pensacola area, and is the designated command for disaster control and hurricane procedures. TraWingSix is unique in that it is the only one of five air wings throughout the Naval Air Training Command training both Student Naval Pilots and Student Naval Flight Officers. Additionally, the Wing is responsible for training a large number of International students from such countries as Saudi Arabia, Italy, Spain, France, Denmark, Sweden, and Germany.

The future of Naval Aviation is determined at the Naval Aviation Schools Command. Within this organization, tomorrow's members take the first steps in their careers. Planning is vital. Every facet must be considered.

The Schools Command prepares officer candidates for commissioned status and provides both aviation indoctrination and ground training to student officers, officer candidates, and naval air crewman trainees. Schools Command also provides specialized indoctrination programs for midshipmen, chief warrant officers, limited duty officers, flight surgeons, future aviation officers, and foreign students. At the present time about 13,000 students are trained annually.

Naval Aviation Schools Command is comprised of four schools: Officer Candidate, Aviation Enlisted Aircrew Training, Officer Training, and Aviation Training. The Naval Air Training Command Choir and the Pageant of Flags are also assigned to the Command.

The Naval Aerospace Medical Research Laboratory is one of the premier research facilities for the causes and cures for disorientation sickness. The primary responsibility of the research laboratory is to conduct research, test, and evaluate aviation medicine and allied sciences. The laboratory staff includes many internationally recognized research scientists, known for their work in vestibular physiology.

The Naval Aerospace and Operation Medical Institute (NAMI) provides professional and technical support and consultative services in operationally related Fleet and Fleet Marine Force medical matters worldwide. It is best known for its training programs leading to designation as a Naval Flight Surgeon, Aerospace Physiologist, Aerospace Experimental Psychologist, Aerospace Medicine Technician, or Aerospace Physiologist Technician. NAMI also offers a three-year residency in Aerospace Medicine and sponsors a course of instruction for Medical Department officers reporting to first tours with ground units of the Fleet Marine Force. The institute conducts approximately 12,000 physical examinations each year for aviation personnel.

The Land Survival Exhibit in Pensacola teaches survival skills to future aviators, flight officers, and air crewmen. The plants and animals on display offer trainees the opportunity to hone survival skills in a non-hostile environment. This exhibit is open to the public Tuesday - Sunday.

One of the newest additions to NAS Pensacola is the Air Force's Water Survival Training Unit, 17th Training Squadron. This joint service effort between the Navy and the Air Force trains aircrew to survive over-water ejections. The squadron works in conjunction with the Navy Water Survival Training to enhance training and save precious tax dollars.

Enlisted personnel are no longer accepted for pilot training. During WWII these men filled a void and provided extremely valuable and honorable service to our nation. However, the Navy offers programs for highly qualified enlisted personnel to earn a college degree and gain entrance into flight training.

Naval Air Technical Training Center supports over 4,000 enlisted students. The center's mission is to train selected Navy and Marine Corps aviation personnel in non-pilot aeronautical technical phases of naval aviation and related subjects. The Center includes classrooms, training hangars, a galley, and barracks for all the students. Each of the 20 schools teaches aviation related skills from aviation electrician to parachute rigger. The school also teaches Army, Air Force, and Allied military forces.

The changes scheduled to take place in the next few years will allow NAS Pensacola and NAS Whiting to develop further as the center piece for naval air training for both officers and enlisted, and continue the strong traditions that began in 1914, to train the best aviators in the world, and most importantly, to be ready to engage in the protection of our country from all enemies. And as NAS Pensacola and NAS Whiting continue to grow, may they carry on the solid friendship with the city of Pensacola and the town of Milton. The assets of the symbiotic relationship between the civilian and the Naval complements must continue in its superb fashion.

As our country calls for change, NAS Pensacola, NAS Whiting, and the entire Naval Air Training Command will jump on board. Never again should this country let its guard down where we must fight our way out of chaos just to get an even start.

CHAPTER 3

Naval Aviation Training Invades Texas

In 1938, after a livid debate, the 75th Congress agreed upon the need to expand Naval Aviation training. Opponents argued that because the Allies had crushed Germany in 1917, the former Axis would not want to engage in the probability of another disastrous defeat.

On the other hand, many congressional leaders proposed expansion on the basis that rogues, even when defeated, will lie in wait for their chance for revenge. They believed we must prepare because Germany appeared to be rearing its ugly head again. This time they were accompanied by Japan and the two of them were becoming more than a nuisance in the free world. They were certain that after being downsized, Pensacola would not have adequate facilities and space to train enough pilots to meet the needs quickly enough in case of a major world conflict. The congressional proponents recommended that Naval Air training be spread to a second front.

Corpus Christi, Texas was deemed as perfect. The Texas area was blessed with wide-open spaces, remindful of a song that soon would be composed and sung, "Don't Fence Me In." An adequate work force was available. Facilities could be constructed quickly with reasonable costs. The expansion proponents won.

Corpus, specifically Corpus Christi Bay, was selected as command headquarters. Captain Alva Berhard received the nod as its first Commanding Officer. The first training flight lifted off on 5 May 1941 in a Stearman, the beautiful and historical bi-winged N3N. Very soon, over 800 instructors had been assembled to train the 300 students that arrived each month.

Ironically, seven months later, 7 December 1941, Japan pulled its sneak attack on Pearl Harbor and the country's leaders were proven to be correct and dead on the money of the expansion question.

The training rate doubled. Flight training reached a furor. In addition to the N3N, students were being trained in planes such Navy SNJ/T6 single-engine trainer (also known in the Army Air Corps as the BT-13), the SNV Valiant, the SNB multi-engine trainer, the OS2U Kingfisher, and the PBY Catalina reconnaissance and patrol aircraft. By 1944, Corpus Christi had overtaken Pensacola and became the largest Naval Aviation training facility in the world. On its 20,000 acres, the facility quickly constructed 997 hangars, shops, barracks, warehouses, educational sites and other miscellaneous structures.

Corpus Christi continued to be an even larger center of a huge complex of Naval training areas with the addition of outlying training fields such as Beeville, Rodd, Cabaniss, Cuddihy, Kingsville, Waldron, and Chase. By the end of World War II, Corpus Christi had turned out over 35,000 Naval Aviators, thanks to the expansion proponents.

Of special note, former President George Herbert Walker Bush (Bush '41) graduated in the third graduating class of June 1943 and was the youngest pilot ever to graduate from Naval Aviation Training at Corpus Christi. Corpus Christi became the home of the Blue Angels from 1951 through 1954. In 1960, Corpus also housed one of Project Mercury's tracking stations.

In the 1950s, Corpus Christi adjusted to the jet age and became the primary site for training Naval, Marine, and Coast Guard jet pilots. In this structure, Primary and Basic training were accomplished in the Pensacola fields of Saufley and Whiting in North Florida. Jet and multi-engine pipeline students were sent to Texas for approximately a year of advanced training.

Advanced students selected for the helicopter pipeline students received advanced instrument training in the SNB at North Whiting Field for three months and then were transferred to Ellyson Field for rotary-wing training for about four months.

With almost certain expected involvement in Vietnam, the Marine Corps directed a large number of advanced Marine students, including many who had been destined for jet training, to switch to Ellyson to satisfy the likelihood of the need for pilots in a brand new type of air war.

Jet and multi-engine pipeline students received basic and advanced training at Corpus Christi in Training Air Wing FOUR — primary to basic through advanced. Because current day jet aircraft are much more complex, the training period was extended accordingly to well over 18 months. TraWing FOUR currently graduates about 400 aviators each year, ready to serve the country.

At the present time, TraWing Four consists of four squadrons: VT-27 and VT-28 handle primary training in the single-engine T34C. VT-31 carries out advanced training in the T-44A and T-44C Pegasus aircraft. VT-35 teaches the twin-engine TC-12B Huron. VT-31 and VT-35 also train pilots from the Marine Corps, Coast Guard and Air Force.

Corpus Christi is more than a flying base. It also is home of the Corpus Christi Army Depot, the largest helicopter repair facility in the world. Corpus also houses a U.S. Coast Guard Air Station and a Naval Aviation Weather Facility.

We continue to offer a huge thank you to the expansion proponents who, back in 1938, foresaw the future.

CHAPTER 4

Whiting and Saufley Training Fields

Whiting Field

Military bases often honor men and women who have served with great distinction and contributed much, but from whom little is asked. This is especially true in the U.S. Naval Services. NAS Whiting Field is an example. No naval officer is better qualified to be the namesake of a naval base than is Captain Kenneth Whiting.

Kenneth Whiting was born in Stockbridge, Mass., on 22 July 1881. He was appointed an Annapolis Naval Cadet on 7 Sep 1900. Upon graduation on 25 February 1905, he was commissioned an ensign. In his heart at that time, his life was to be on, in, or around the sea. As was legally required, he began his service on the sea.

It was not that surface duty was boring, but Whiting wanted a more intense course of service. He volunteered for submarine duty. The Navy consented, trained him well, and put him in command of the Porpoise, the Tarpon, and the Seal. In 1909, one of his most interesting accomplishments came to light after his study of how submariners, trapped in a disabled sub, could be saved. In Manila Bay, he experimented by submerging his submarine to a depth of 20 feet and, in a simulated emergency,

Whiting himself swam out of the 18" torpedo tube to what would have been a successful rescue.

A year later, in 1910, civilian aviator Glenn Curtiss became interested in getting into the aircraft business and offered to train men to be naval aviators. Whiting heard of this program and applied. At the same time he suggested to his friend, Theodore G. "Spuds" Ellyson, that he should do the same. Unfortunately for Whiting, Ellyson was accepted over him and eventually became Naval Aviator #1.

In 1914, flying activity picked up. Whiting reapplied, was accepted, and became the last navy flyer to train under the teach yourself to fly syllabus and was designated Naval Aviator #16.

Whiting, still possessed of creative ideas, developed the idea of the aeroplane carrier, as it was talked about at the time. He converted a Naval vessel, the Jupiter, into the Navy's first full-fledged aircraft carrier, renamed it the Langley, and became its commanding officer. With this success on his plate, he began to lobby for carriers as part of the Fleet.

Captain Whiting assumed command of the Naval Air Station, New York in the early 1940s and remained there until his death on 24 April 1943. NAS Whiting Field, Milton, Florida was rightly named in his honor.

NAS Whiting Field has become the principal training location for Naval Flight Training, incorporating the training that at one time was split between Saufley, Whiting, and Ellyson. Whiting and Saufley Field are included in this section together because up to a point in Naval Aviation Training, the two locations could have almost been considered twins even though physically they are forty miles apart. Both fields went into operation in the 1940s as auxiliary fields (NAAS). In 1976, Saufley was redesignated an outlying station, and continues to serve the Navy administratively as will be seen later in this section. Whiting assumed the lead and has become the largest and most active flight training station in the United States Navy.

In 1945, Whiting Field also housed World War II German prisoners. Many of these prisoners were offered the opportunity to join work details aboard the military bases and even to work outside the bases on civilian farms. The work teams of that era built several of the older buildings that remain in use today. Prisoners received a fee for their services, certainly better treatment than what Allied prisoners received in German and Japanese prisoner of war camps.

This author remembers World War II from the eyes of a child. German and Italian prisoners were housed at a stockade near Augusta, Georgia at the U.S. Army's Camp Gordon (now Fort Gordon):

> *I was about eight years old when my father and grandfather needed workers to harvest crops and work in our dairy business. Most of our regular helpers had either enlisted or had been drafted into the military. Other farmers were in the same boat. Within our community, farmers always worked together and shared workers. But now, with many of these employees off fighting in Germany and the South Pacific, agricultural labor was difficult to find.*
>
> *My dad and granddad arranged with the Army for us to use willing prisoners of war. The prisoners seemed to enjoy this and it was within the rules of the Geneva Convention. Their assistance was voluntary and they were paid. Sometimes our farm was provided Italian prisoners and other times German. Interestingly, Germans and Italians were rarely placed in the same field. We didn't know why but we assumed that Germans and Italians didn't get along well. We complied because we wanted no trouble.*
>
> *We were requested not to mingle with prisoners. They called it fraternization. However, when the prisoners were away from the Army base and safely in our fields, some guards would change the rules.*
>
> *We kids looked forward to days prisoners would arrive. We would be permitted to chat with them before work and during breaks. We would carry water and snacks to them. They reciprocated by showing us items from their country. They were especially interested in showing us family pictures. During breaks, they would*

spread these pictures on the ground and talk about their kids. It was not until I grew older and remembered how happy they were to share their pictures that I realized how lonely they must have been.

Prisoners were usually alternated from farm to farm but often the same prisoners would come to our farm. I especially remember an Italian named Tony. One late fall day, a host of prisoners arrived and I wanted to meet all of them. Then I saw my friend, Tony. He waved and ran up to me. I called the guard to help with our language.

I was wearing a little black cloth coat with a bright silver zipper. Tony wanted to feel it and to touch it. Of course I agreed. His face brightened as his hands ran up and down the slide and the teeth of the zipper as he watched the interlocking teeth engage perfectly. Tony and I, with the guard as interpreter, talked about the zipper. Tony had never seen a zipper, so he took out his little notebook, sketched a drawing of it and wrote something.

Tony became my friend. I remember saying, "Bye, bye, Tony," each day after work as the truck pulled away. I have often wondered, if when he returned to Italy, he might have started a zipper company, named it Talon-a-Tony Zippers, and made a fortune exporting these to the U.S. More power to you, Tony.

Whiting Field, since those Forties, has served the Training Command well. It has been home for medium and heavy bombers, home for the famed Blue Angel flight demonstration team, and a jet training base. Since 1965, it has been the site where most Naval flight students receive their very first Naval flight hours. Whiting is also the only facility that trains helicopter pilots for the Navy, the Marines, and the Coast Guard.

Whiting, at this writing, houses two complex flight simulators, a night vision goggle laboratory, many training classrooms, more than 275 aircraft and 15 outlying practice landing fields. More than 160,000 flight hours are flown each year at the station, making it the most active Naval air facility in the country. The air station boasts over a million and a quarter flights each year.

A bit of a statistic as this book is written: One of every two Naval Aviation flight hours (all classes) across the entire globe is flown out of NAS Whiting Field. Twenty four runways totaling 800,000 linear feet (slightly more than 150 miles) are divided between North Whiting and South Whiting Fields, and are within a mere football field and a half of each other. The flight patterns of the two fields are so creatively and cleverly designed so that even in times of very heavy take off and landing volume, day or night, traffic from either field is not distracted by the activity on the other field. Three aircraft traffic control towers monitor the flight traffic 24 hours per day.

The entire Whiting complex covers more than 12,000 acres of owned and leased land, including 2350 acres of woodlands. Just as NAS Pensacola has maintained an elevated relationship with the city of Pensacola for a hundred years, Whiting enjoys the same enviable rapport with the citizens of Milton. On days of military celebration, such as Armed Forces Day, the gates are opened to civilians and they are treated to magnificent displays of Naval Aviation. Students from more than a dozen Junior Reserve Officers Training Course units from across the area visit Whiting each year and receive royal tours of the entire facility. Today, all six training squadrons of Training Air Wing FIVE participate in the *Pilot for a Day* program, helping seriously ill children enjoy a special day on the facility.

Saufley Field

Joseph Trawick remembers Saufley:

I remember flight school in 1958. The stuff in Preflight was new to me and I was thankful to graduate as a NavCad with average grades. Nervously I drove the ten miles down Saufley Road and soon saw all those little yellow planes on the tarmac. I looked up and saw a huge sign that assured me I was approaching "Saufley Field." Here I would break away from the earth and actually fly an airplane. I was almost to Primary Stage and I had never ridden in a plane. I didn't know whether I would like it. I didn't know if I could hack it.

Here was the stage our Preflight instructors told us about. We would climb into those tight-fitting cockpits and get our first taste of Navy flying. Navy flying? Crazy. It would be my first time for any kind of flying. But I kept my mouth shut about that. I didn't want anybody to know.

I thought about Saufley being my launching pad to Whiting Field. Whiting! The idea of Whiting and huge T-28 aircraft gave me trembles. This was crazy. I hadn't even touched a T-34 yet and here I was, dreading the T-28. I wondered if I had done the right thing. Maybe my mama was right. Maybe this is too far above me.

Then I realized I had to get my good hat on. I had whipped Preflight. I learned what I had to learn up until now. Being this far along meant that if I just kept going I would soon be standing out front of the gates of Whiting Field. I had to get there. I had promised my family I would. If not, I'd be joining the others who failed and would leave the beautiful Florida Gulf Coast. I didn't want to be a supply officer or be stationed ashore as a pay officer. I had promised my family I would make it. Mainly, I promised myself.

The end of this story is that I did it. I got home, opened the door, and walked inside sporting the uniform of an ensign and the pretty wings of a Naval Aviator. The first thing my mama said was, "Joe, I knew you could do it." Mamas are like that.

The world situation heated dramatically toward the end of the 1930s. Nazi Germany and Japan began to run rampant in their pursuit to gain territories, countries and even the world. In a move to counter the exploits of these two powers, Congress passed the Naval Expansion Act in 1938. Three new Naval Air Stations; Jacksonville, Miami, and Corpus Christi, TX; were approved to support the Navy's growing flight training program. In addition, the Navy included a fifty percent expansion of Pensacola's facilities.

Once this expansion was begun, the Navy purchased 869 acres of land, called Farm Field. Farm Field had been leased by the Navy since 1933 and used as an outlying aircraft practice field for Corry and Chevalier Fields. Construction started in August of 1939 at the site of a field soon to be named for LTJG Richard Saufley, Naval Aviator number 14 who was killed in a plane crash at Pensacola on 9 June 1916.

An interesting side note on Farm Field — when the Navy bought Farm Field in 1939, the land that would soon become Saufley Field, an older couple in their eighties occupied a small residence on the property. The Navy allowed them to keep their home and even built a special access gate for them to come and go as they pleased. The Navy also provided them with medical care, food, and a handy man to keep their property neat. The gentleman died in 1943. This arrangement continued until 1948 when the lady of the house passed away.

Let this tidbit be told to those groups of people who today live under the protection of our armed forces, but argue that the "appalling military has no heart".

Saufley Receives Training Squadrons

Saufley Field opened in August of 1940 as an auxiliary practice field. The aircraft parking area soon boasted of 50 North American NJ and SNJ Texans and proudly flew the flag of instrument training squadron, VN-5. That squadron was redesignated as VN-3 four months later. Then three months later, Stearman NSs, N2S Kaydets, and N3Ns arrived. Saufley was officially commissioned as a Naval Auxiliary Air Station (NAAS) on 1 March 1943.

The Training Command began to grow so swiftly that the number of personnel in squadrons also grew quickly. The newly-designated VN-3 was split with both squadrons having the same primary designation but with a different letter appended to each. For example, when VN-3 was divided, the result was two squadrons: VN-3A and VN-3B.

Saufley became a very important part of the wheel of the Training Command through the chaotic years of the Second World War. The field was positioned well clear of Pensacola and could be used by the Navy with little concern about disturbing the civilian population. In November of 1941, just before the beginning of World War II, VN-3 moved its instrument training facilities to the outlying Chevalier Field to eliminate dangerous sky clutter.

Early in 1942, the Navy modified its training plan and directed that primary stage flight training be spread over the country using "Yellow Peril" Stearman trainers. NAS Pensacola and NAS Corpus Christi were selected to handle all basic stage flight training. Operational training moved under control of NAS Jacksonville. Primary training squadrons VN-1C and VN-1D were closed in January 1943 as their duties shifted to other locations.

With the Training Command in constant limbo trying to bring together the most workable solution, Saufley lost its training mission. Most of the brand new Naval Aviation students had not yet warmed the seat of any aircraft. Basic Stage students, those who had attained as many as 40-50 hours in their logbooks and maybe felt that they were ready to meet the foe, flew out of Old Farm Field in the Vultee SNV Vibrators. Then, instrument training returned to Saufley as squadron VN-3B arrived. However, VN-3A was shifted from Chevalier to the newly opened NAAS Whiting Field, and with it, VN-3B joined its sister unit there.

Continuing to play fruit basket turnover to the hilt, the Training Command sent VN-2B to Saufley in May. VN-2B began to conduct training in the Consolidated PBY-5A Catalina for Army Air Force pilots and flight engineers as they trained for air-sea rescue squadrons.

Basic Stage flight training at Saufley ended again on 28 September 1944 when VN-2B was disestablished. Taking VN-2B's place was VN-9, which was responsible for the newly established pre-operational training program for all carrier bound students. This course consisted of three weeks and 26 flight hours of familiarization, formation, navigation, and ordnance training in the 150 Douglas SBD-5s on the field. On 15 December 1945, VN-9 was redesignated VN-5 and assigned the mission of carrier qualification. The SBDs remained at Saufley until the spring of 1946 then retired.

In November of 1945, the Training Command was split into the Basic Stage Training Command and the Advanced Stage Training Command and placed under the command of the Chief of Naval Air Training. Primary and Basic Stage instructional classes were conducted at Corpus Christi. Multi-engine land and sea carrier training (included carrier qualification) was conducted in the SNB, the PBM and the SNJ.

Saufley escaped shutdown after World War II, primarily because its facilities were so well constructed. Its all-weather operating capability was up to date. At the end of 1945, the Carrier Qualification Training Unit from Glenview transferred to Saufley. With it came the SNJs forming VN-6 and by December the instructors were teaching basic flight to students in SNJs and carrier qualification in the SNJ-Cs. VN-6 was now the basic carrier qualification training squadron of the Navy. By June 1947, an advanced squadron, CQTU-4, was formed by bringing in the F6F, the SBDs, the TBM, and the F4U for carrier qualification. In May of 1949, the squadron moved to NAAS Corry Field.

During the last half of 1947, the Stearman was phased out of the Training Command. SNJs and other single-engine instrument aircraft were brought from Corpus Christi to Pensacola. All multi-engine and advanced training moved to Texas. Still in fruit basket turnover mode, VN-5 became BTU-3 and taught basic formation and tactics. In 1953, BTU-3 was redesignated BTU-2 and performed the same mission as before.

As the Korean War ended, the Navy introduced two superb new trainers to replace the old SNJ workhorse. In 1955, the T34B Mentor, built by Beech Aircraft, arrived at the Whiting Field Primary Training squadron in BTG-1. In 1957, the North American T-28 Trojans were delivered to the Basic Training squadrons at Corry and Saufley.

In October through December of 1956, more fruit basket moves occurred when BTG-1's Primary Stage operations moved from Whiting to Saufley and all T-28 Basic Stage operations transferred to Whiting. In November of 1958, the CQ basic training field, NAAS Barin, closed, sending BTU-5 to Saufley with its T-28C trainers.

New Aircraft and Changing Tasks

In July of 1958, BTG-1 received its first Temco TT-1 Pinto jet trainer. Jet, yes. Mini, yes. But it soon received the name, a mini-meeni-jet. At a mere 900 pounds of thrust, the T-28 could outclimb it to 20,000 feet and beyond. Even the T-34B would give it a scare. The purpose of the TT-1 was to serve as the front-end of an all-jet, start to finish, program throughout the Training Command. After 40 hours in the TT-1, successful primary students graduated to the more powerful T2J for Basic Stage jet training. BTG-1 received only 15 of these Pintos. This was a bit short of the count for a training philosophy.

Three years later, in 1960, the Navy disbanded its TT-1 and T2J all-jet program and the TT-1s were retired. At the same time, BTG-1 and BTU-5 were designated VT-1 and VT-5. On 31 July 1968, Saufley was officially designated and named NAS Saufley Field.

In the 1970s, the Training Command underwent still more reorganization, this time to the single-base training wing concept. Training Wing 7 was established at Saufley on 1 February 1972.

Carrier qualification for basic propeller students was eliminated on 15 November 1973. At the same time VT-5 began its transition to a primary training squadron. VT-1 and VT-5 were put under the control of Training Wing Six at NAS Pensacola.

Saufley and its training squadrons were retired. Budget cuts, a horror for the mission of all military training, had come to pass. That was reason one. Reason two was that the more powerful new T-34C was replacing the T-34B and the T-28. The last T-34B flight was flown by VT-1 on 27 September 1976. VT-1 and VT-5 were disestablished on 1 October.

No longer would we hear, "Saufley Tower, Saufley Tower, this is Two Sierra 1-1-6. Radio check. Over." Or another of those calls from some nervous student, maybe trying to delay, if only by minutes, getting into the air on his A13 first solo flight when he would be aloft and alone.

One of the chief goals of primary flight training was, as it is today, to help students become accustomed to aircraft on the ground and in the air. Many of these students have never flown before. And aside from that, quite a few foreign students were in the mix. Language was a constant problem for most.

Max Chambers was a flight student in 1955. He enjoyed talking to the Spanish-speaking students. It wasn't necessarily that he wanted to be a goodwill ambassador. He had studied Spanish in college and thought this would be a good way to test and upgrade his knowledge. Listen to the story that Max remembers. He calls his story **In Zee Chocks**:

> *Foreign students were an interesting segment of our flight student body during my days of flight training in 1955. I enjoyed talking to them. I had studied Spanish in college and thought I could speak the language fairly well. That was before I began to try converse with a Peruvian student in the same Primary Stage flight squadron as I.*
>
> *What a shock. I found that I wasn't as well educated as I thought. Others speaking languages just as difficult came from Italy, Spain, Japan and South Korea. Hearing them talk to each other brought me to realize how much*

of an obstacle language differences must be. I could hear the tower spurting words I instantly recognized, but must have been trying to their ears. We did, however, have students from England, Australia and New Zealand. Except for accents, communicating was easy.

Most of the foreign students were probably handpicked by their country and were a bit higher on the ladder of knowledge. Even so, there is no doubt that they had to struggle with learning to fly as well as to overcome the language barrier as well.

One day at Saufley, the one o'clock launch busied itself for takeoff. Several planes had already departed and several more were jockeying for position to take the duty runway. About every twenty seconds another T-34B lifted from the ground. I had just fallen into the back of the line waiting and listening to the tower trying to do their job as efficiently as possible with these inexperienced students.

Suddenly the air was electrified by a transmission over the emergency channel from my friend, the Peruvian, "Mayday, Mayday, Saufley. I have zee engine failure."

The tower operator responded quickly and calmly, "Aircraft with engine failure, what is your position?"

The answer remains in my memory as strongly as if I had heard it this morning, "Rogg-air, Saufley. I am in zee chocks. Zee engine is dead. I am waving zee handkerchief out of zee cockpit."

Primary flight training shifted to Whiting and Corpus Christi. Saufley Field was closed and disestablished on 1 December 1976 as a naval air station, but remained opened as a day VFR outlying field for practice landing by Whiting-based aircraft.

However, it was reactivated in May 1979, not for flight training, but to house the Naval Education Training Program Development Center as it moved from the long-time helicopter training base at Ellyson Field.

Today, much to the sorrow of many former Naval Aviation students, Saufley receives aircraft no longer. It is home to a minimum security Federal prison camp, a Naval Reserve Center, CNET's Professional and Technology Center, and several education and financial systems centers. Why did they do it? The memory of Saufley is like our first date, the first fish we caught, the day we learned to swim.

A Reminiscence of Saufley

This writer has very fond memories of Saufley Field. That was where I first flew. Many former Naval Flight students of my era (late fifties and sixties), friends, and those I have since met, recall with absolute delight most of our days in flight school. The T-34B was a magnificent first aircraft. It soothed, lightened, and enlightened. It didn't take long for the T-34B cockpit to become our home; to become our safe-house.

Saufley Field cannot escape being the object of fond memories. It was the first leap into flying the Navy way for a few; the first leap of any sort of flying for most.

The consensus of former students I interviewed is that the Navy should never have retired the T-34B nor closed Saufley Field, if for no other reason, because of our deeply seeded, and I must admit, selfish and emotional memories. Today, as you drive out to Saufley Field west on — you remember – Saufley Field Road, now State Road 296 — it is easy to remember those yellow birds sitting proudly in the chocks, pleading to Primary Stage students, "Take me! Take me!"

CHAPTER 5

NATOPS

The Naval Air Training and Operating Procedures Standardization (NATOPS) program standardizes general flight and operating instructions and procedures applicable to the operation of all US naval aircraft and related activities. The program issues policy and procedural guidance of the Chief of Naval Operations; applicable to all Navy, Marine Corps, and Coast Guard aviation personnel.

What is NATOPS

NATOPS is an approach toward improving combat readiness and achieving a substantial reduction in the aircraft accident rate. Its basis is standardization throughout the Navy, based on professional knowledge and experience. The standardization program is not designed to stifle individual initiative, but rather to aid the commanding officer in increasing a unit's combat potential without reducing command prestige or responsibility.

Why NATOPS

NATOPS was created by the U.S. Navy in 1961 to improve combat readiness and achieve a substantial reduction in naval aircraft mishaps. In 1950, two aircraft per day (776) were lost in 54 major mishaps per 10,000 flight hours. Several major initiatives were credited with significantly reducing the rate of mishaps from 54 to 19 by the year 1961. One of the most significant, in 1954, was the angled flight decks of carriers. By 1970 this rate had decreased to nine. Currently the rate is under two major mishaps per 10,000 flight hours. Why not NATOPS?

This improvement deserves a bit of emphasis: For the 20 years between 1950 and 1970, the accident rate for 10,000 hours of flight time was reduced from 54 to two. NATOPS deserves a pat on the back.

The standardization initiative began in 1961 with the introduction of the Fleet Replacement Squadron program. FRS indoctrinates newly designated aircrew and aircraft mechanics into the peculiarities of specific aircraft. Prior to the FRS concept, qualified pilots transitioning to a new aircraft were essentially told how to start it, and then sent to go fly.

NATOPS Implementation

NATOPS manuals contain standard flight doctrine and the optimum operating procedures for the aircraft model or aviation activity concerned. They do not include tactical doctrine. There are numerous publications associated with NATOPS covering several basic areas. These manuals are available in every flying unit and upgraded upon any significant change.

NATOPS is not intended to cover every contingency that may arise nor every rule of safety and good practice. However, aviation personnel are expected to study and understand all applicable portions of the program.

NATOPS Flight Manuals are prepared using a concept that provides the aircrew with information for operation of the aircraft, but detailed operation is not provided. This concept was selected because reader interest increases as the size of a technical publication decreases.

KEY PEOPLE in NATOPS range from the Chief of Naval Operations all the way down to individual users.

EVALUATION is performed annually using an open book examination, a closed book examination, oral examination, and an evaluation flight/simulator. Use of trainers is encouraged for simulated emergencies and scenarios that present significantly increased risk when performed in an aircraft. If no such device is available, aircraft may be used. Evaluation flights in aircraft that require simulated emergencies are avoided.

Only one CONCLUSION can be made if we refer to the results shown after over four decades of the use of NATOPS. It works. No other conclusion may be reached.

CHAPTER 6

Aviation Safety and Survival Devices

Since 1903, the banner that moved flying from a pipe dream status to reality is called *pilots*. These are men and women who flocked to these strange machines, learned them, and married them. Along with that distinction, these pilots faced an inherent danger — what happens when my machine breaks? The military, private aviation, and commercial aviation, constantly develop procedures and devices designed to save lives of personnel and to conduct operations more effectively and efficiently.

Parachute

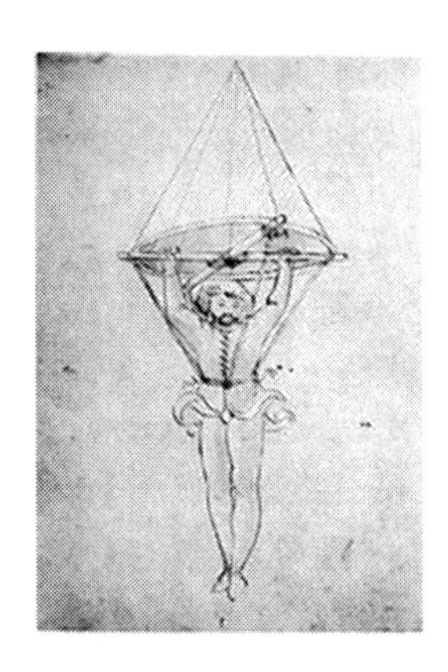

One of the earliest safety devices pairing the pilot and the aircraft for safety was the parachute. It would be impossible to count the number of flyers' lives it has saved. When an airplane engine fails, or an aircraft is downed by enemy fire, the parachute is the pilot's primary means of survival. In addition to saving lives, the parachute has contributed to the success of many military operations when fighting units have been dropped far behind enemy lines.

Many of us might think that the parachute is a relatively recent invention. That is far from the case. The very oldest record of the idea of the parachute comes from Renaissance Italy in the 1470s in the form of a drawing. This drawing shows a human descending as he holds onto a cross bar attached to a conical canopy filled with air. The canopy seems to be no more than six to eight feet tall with a diameter of about four feet. Those of us with a remote understanding of aerodynamics can look at that picture and realize that that particular design is virtually unfeasible but the concept is real.

Less than 20 years later, in about 1470, Leonardo da Vinci, the most notable of all inventors, published a parachute design with a much more favorable proportion. Even with his creative designs, the parachute had yet to be flight-tested.

In 1783, a Frenchman, Louis-Sebastian Lenormand proved the parachute when he made the first actual parachute jump from the tower at the Montpellier Observatory. Finally in 1911, a bona fide use of a parachute was realized. Grant Morton jumped from a Wright Brothers Model B aircraft flown by Phil Parmalee at Venice Beach, California. Morton stood on the wing and held the chute tightly clasped in his arms. Once the plane reached several hundred feet of altitude, Parmalee jumped. Once he cleared the plane, he tossed the chute into the slipstream. The rushing air filled the parachute canopy, snatched Parmalee from the wing, slowed his descent, and eased Parmalee to the ground.

That same year, in 1911, a Russian, Gleb Kotelnikov, invented a knapsack parachute. The canopy of the chute was encased in a knapsack. When the parachutist cleared the plane, he pulled a ripcord that opened the knapsack and pulled the canopy into the air. With this model, the assurance that the pilot would reach the ground safely greatly increased.

During World War I, Germany pioneered the development of parachutes in combat aircraft. These devices were less than efficient but did save lives. All countries using aircraft began to employ chutes to save lives. Efficiency increased. In World War

II, not only were almost all aircraft equipped with parachutes, but as an additional tactical scheme, ground troops were trained to jump and huge numbers could be deployed quickly and efficiently by having them jump into battle zones — even far to the rear. This was most notable during the Normandy invasion in 1944 when Allied troops were dropped in large scale far behind the lines.

Many nations began to look toward space. The United States and the Soviet Union each realized that the other had its eye pointed up. It was no doubt that each knew they must latch onto the theory stating that the nation controlling space controls the world.

American Colonel Joseph Kittinger set the record for the highest altitude parachute jump. This test was conducted as a prelude to the aerospace program in which the U.S. was a leader.

Strangely enough, Kittinger's jump platform was not an airplane. He climbed aboard a gondola attached to a gas balloon on August 16, 1960 and ascended to an altitude of 102,800 feet. He literally leapt into space by plunging over the side. Aided by a stabilizer drogue chute, he fell free for 4 minutes and 36 seconds before deploying his chute. To get a full understanding of that free fall, the reader should look at a watch and wait while the second hand rotates for 4 minutes and 36 seconds — 276 seconds.

After the parachute opened, Kittinger continued to descend, under control, for another 13 minutes and 43 seconds. During the free fall phase, he reached the expected terminal velocity of a human — about 120 miles per hour. For the entire 19 minutes of descent, Kittinger survived high altitude by the use of a highly efficient flight suit and a small, slender oxygen canister connected to his oxygen system.

The colonel had not yet finished his association with parachutes. Later, during the Vietnam War Colonel Kittinger was serving as a fighter pilot in the 432nd Tactical Reconnaissance

Wing on his 483rd combat mission. Over Thai Nguyen, North Vietnam, in May 1972, his plane encountered a flight of MiG-21 fighters and was shot down. Kittinger and his Weapons System Officer, 1st Lt William Reich, ejected, were captured and spent the rest of the war in the Hanoi Hilton.

Ejection Seats

Everard Calthrop was a close friend of Charles Rolls, the principal owner of the Rolls-Royce automobile company. Rolls was quite a figure in aviation and was the first man to cross the English Channel and back again, roundtrip by air. On the 12th of July 1910 Calthrop accompanied Rolls to the Bournemouth International Aviation Meeting, and was present when Rolls died after losing control of his biplane and crashed. That and a similar, non-fatal, accident involving his son, Tev, led Calthrop to become almost certain that a parachute is the answer to saving pilots in similar circumstances.

In 1913 he patented his first parachute. As World War I came closer, he continued to make improvements to his parachute. In 1915 he offered it to the Royal Flying Corps and successful tests were completed at the time. As naysayers will always do, rumors surfaced with the opinion that parachutes will impair the fighting spirit of pilots, so the offer was rejected. Calthrop was encouraged to remain quiet about his invention. Continually faced with increasing losses of pilots, he publicized the parachute in 1917. Despite a campaign by some pilots, the Royal Flying Corps declined to introduce parachutes during World War One, although the air forces of most other nations did so.

Calthrop's "Guardian Angel" parachute received much praise and was used during the war to drop agents behind enemy lines. By 1918 it was known that the Germans were fully aware of Calthrops work, and supplied their pilots with a similar design. However when the Royal Air Force finally adopted parachutes after the war, they chose an American design.

In 1916, Calthrop invented and patented an *ejector* seat for aircrafts. These devices used compressed air to eject. This device, an extension of the parachute realm, used a charge of compressed air to shoot the airplane pilot from his cockpit to a safe distance from the airplane into the air stream by a charge of compressed air and open his life-saving parachute. The pilot only needed to pull a lever, relieving him of the stress of having to unbuckle his harness, climb out onto the wing and actually leap.

Sweden developed a very effective compressed gas powered system in 1941. Then they followed with a system powered by an explosive cartridge – probably the origin of the phrase, *kick in the butt.* All these early creations should greet aviators with joy, because, if an aircraft experiences an emergency even on the surface of the runway, he may eject with the expectation of a safe landing.

In the early jet fighters, airspeed was so great that clearing the cockpit during an emergency by traditionally jumping would be almost impossible. During the early stages of development, air powered ejection seats had to be triggered above a thousand feet or the pilot would hit the surface before the chute could fully deploy. With constant improvements, the successful altitude dropped lower.

In the middle fifties, the Navy developed its ejection seat trainer. They were first implemented in the fleet, then continued in Naval Flight training. These trainers looked almost like small oil derricks anchored into the ground and bent over about thirty degrees from the vertical. Two narrow parallel rails, about 25 feet long, looking much like the rails on a railroad track, were bolted onto the derrick. A seat was fastened to a carriage with four wheels at the base of the derrick. The wheels on the carriage would ride the rails, attaining an altitude of about 20 feet, then settle again to the bottom. Most of these first Navy trainers were set up out on the tarmac where the students were at the mercy of the weather.

Low-pressure Chamber

Aircraft science developed. Planes flew faster and higher. They soon could fly high enough so that the occupants reached such altitudes that oxygen concentration was so low it could not support the requirement of the human body. Something was needed for pilots to take oxygen aloft case of an emergency.

The oxygen mask was born. This device worked in an aircraft until something went awry. Something going wrong might be gunfire from enemy planes or antiaircraft fire that would penetrate the plane's oxygen supply and cause a rapid loss of oxygen. It might mean an unnoticed fault in the oxygen system.

Pilots had to be able to recognize hypoxic conditions to quickly make corrections and descend to lower altitudes or to bring into play devices that counteract this hypoxia. And these decisions had to be made quickly. The low-pressure chamber, sometimes referred to as the high altitude chamber, was invented. Its purposes were (1) to teach airmen to recognize oxygen deficiency and (2) to learn to take action.

Naval Aviation pilots and air crewman are required to experience hypoxic conditions by visits to the low-pressure chamber many times. Flight students first encounter the chamber during preflight at Pensacola before even stepping into an aircraft. In this event, students listen to a thorough indoctrination. This seven-foot high by twelve-feet long round chamber demonstrates to students how this knowledge may one day save their lives.

The students are monitored every second by trained observers inside and outside the chamber. Aerospace Physiologists and Technicians are available in an instant if emergencies or malfunctions occur. A hyperbaric chamber with a trained team is posted outside, ready to act in case of an emergency.

In 1993, LCDR Michael Venable was a Naval Aviator stationed at the Naval Aviation Flight School. One particular day, Mike was assigned to monitor two Aerospace Physiology Technicians during a 25,000 feet low-pressure chamber exercise for a preflight indoctrination. Michael would be working inside the chamber to monitor a group of students.

> *I was assigned to work as an inside observer for a class of preflight students in Pensacola. As the students filed into the amphitheater area, they were issued oxygen mask packets and presented with the normal lengthy*

indoctrination. One of the major points stressed was that during the event, students were not to anticipate commands nor change settings on their paraphernalia. One by one, technicians answered questions about the equipment and hoped that the information they received would calm their uneasiness.

First they asked if any student had experienced previous medical issues. No students answered. The chief trainer then spoke for a few moments and was satisfied.

The chamber, positioned off to the side, waited its turn for the students to climb aboard. It was obvious the class was a bit restless. Every class does. The door to the chamber was opened and everyone climbed in. Nervousness reached an even higher level. The chief trainer pointed out the gauges showing the pseudo-altitude and the actual pressure inside. It showed a correct pressure of 14.7 PSI, the standard barometric pressure at sea level.

The event began. One of the assistants inflated a medical rubber glove to resemble a man's hand and tied a knot at the wrist. He suspended the rubber glove about six inches under a red hook at the top of the chamber. With everyone inside, the hatch was locked and bolted shut with an ominous sound, one similar to the sounds we hear in old submarine movies as crews close the hatches as they ready for battle.

The chief trainer gave a hand salute through the observation window to the crew outside and a rather eerie air pump sound started. The nervous and edgy chatter dropped to almost a murmur. Students seemed to direct their stares onto the altitude gauge, which by now had climbed past 1000 feet. No problem. In fact, the glove had hardly changed size.

At 5000 feet, the glove looked puffy, as if a honeybee had stung a human hand and it has started to swell. The pressure showed 12.25 PSI, a bit less than the normal mean sea level pressure of 14.7. The pump continued to evacuate air.

A few minutes later the altitude gauge registered 10,000 feet, with pressure now down to 10.1 PSI. The appearance of the glove had now changed from honeybee sting size to how a human hand would look as if attacked by a couple of hornets. The glove had doubled in size, but the class still had 15,000 feet to go until reaching its goal. The faithful pump kept running, exhausting the chamber of life-salvaging oxygen.

After another several minutes, the glove had reached three times its original size and the altitude gauge read 15,000 feet. The pressure gauge inside the chamber had dropped to 8.3 PSI; over half way. Conversation among the students had lulled so that only a couple of them were wisecracking, even though nervously. The chief trainer winked at me.

At 20,000 feet, the pressure gauge displayed 6.75 PSI and the balloon was fully four times its size when we began. Students began staring out of the observation window, probably anxious to see that none of the staff seemed worried. Only 5000 feet remained until we would reach our goal altitude of 25,000 feet.

Just before we reached our goal, one of the students tugged at my flight suit. I turned around quickly and he pointed to another student, appearing almost lifeless and very pale. I notified the chief trainer. They stopped the ascension and immediately set the chamber to begin a descent at a safe 5000 feet per minute.

The corpsman inside took over and tried to revive the student but was unsuccessful. He set the student's mask to 100% oxygen and ordered the remainder of the group to do the same. I ran my hand inside the student's mask and could feel oxygen rushing in. Finally, at 18,000 feet, some color entered the student's cheeks. But he would not recover more than this.

As we dropped below 10,000 feet the hyperbaric team, including the medical officer of the day, opened the hatch and rushed in. The student remained very slow to recover. He was quickly removed to an adjacent emergency room.

An investigation revealed that several times in his life, the student had fainted or felt woozy when blood was drawn from his body. Everyone breathed a sigh of relief. This meant he was one of the rare people in which some condition of anxiety causes a serious vascular event in the body. This results in a harsh decrease in heart rate. He was referred to Medical and dropped from the flight program. Even though this condition is easily medicated, it absolutely disqualifies a person from flight crew status.

The Dilbert Dunker

The Dilbert Dunker set many a flight student on his ear; probably well before they crossed the Pensacola city limit sign on the way to the Naval Air Station. No doubt the Dunker was known far and wide. It was a scary looking apparatus. The original Dunker was born in 1943 at NAS Pensacola. It was a time of war. Many Naval Aviators and crewman were losing their lives as aircraft ditched and they were on their own, without special training, to survive a water crash.

Army Colonel Wilfred Kaneb was tasked with designing an apparatus that could save airmen lives. Kaneb's device worked. The long ride down the track of that contraption seemed chilling enough, but the splash, turning upside down, then sinking convinces the aviator that he is drowning. This training continued after flight school to the fleet and the Navy attributes thousands of saved lives to Kaneb's invention.

No unusual secret was involved in the Dunker's ability to protect an airman. It was simply a matter of getting an aviator accustomed to unusual attitudes under water. Just as an aircraft might do when ditching, Kaneb's device might flip upside down; it might skip across the water and spinning; it might immediately go straight down.

The student would not loosen his seatbelts until the device had stabilized. To do so earlier might hurl the student violently about, knock him unconscious and unable to function; certainly a killer. Once the device was stable and began to sink, the airman would learn to unlock his harness and seatbelt and swim out of the cockpit to safety.

As with all aircraft, ships, boats, and the like, this Dilbert Dunker also had a technical identification. It is called the 9E8E. Who knows what this designation really means? Without an answer, that will remain an uncertainty.

Since 1943 until recently, every Naval Aviation student has donned his flight suit and parachute, strapped into the 9E8E, has ridden from the top of that scary platform, picked up speed as it descended the dual tracks, slammed into the pool full of water, flipped upside down, and slowly sank into that deep swimming pool. But once the student unfouls his chute straps, extracts himself from the cockpit, and swims away safely, does he believe. Not one life was ever lost in Dunker training. It would be very difficult to estimate the number of lives it has saved.

Most students handled the escape from the Dunker cockpit cleanly. In case of trouble, safety swimmers lingered in the water, ready to assist. And most of all, no flight student ever earned his wings without at least two successful Dunker rides.

The Dunker that trained approximately 8300 Naval Aviation flight students over the years achieved some degree of fame other than being a bit frightful. The very same Dunker that we, as Naval flight students, climbed aboard in our training, was used in two movies: "Flight of the Intruder", Brad Johnson and Danny Glover; and "Officer and a Gentlemen", Richard Gere and Louis Gossett.

The 9E8E Dunker retired on 14 NOV 2003. Navy Captain Stephen Black was selected for its last dive. He stood in for us all in saying thanks to the Dunker that had served us well for 60 years.

As a replacement test vehicle, the Naval Operational Medicine Institute, as of 2004, is designing a new $4.5 million state-of-the-art Aircrew Water Survival Training Facility at Whidbey Island in Washington State.

CHAPTER 7

The Road to Pensacola

The Road to Pensacola is a tough route. But you must take it to be accepted for Naval flight training. It can be lengthy. At the end, it can be gratifying. For some, it may be heartbreaking. Every person will not be accepted. So many factors enter into the equation. (Remember in the preface, the Navy commander's speech to a new preflight class?)

Naval recruiting stations showed pictures of aircraft and magnificent young men and women sporting flashy dress uniforms, ribbons, and golden pilot wings on their chests. As a kid you might have said, "Hey, Momma. That's what I want to be." It probably was at the time. As the months and years rolled by, the pretty pictures might have faded from your mind. On the other hand, the pictures might have stuck, turned into a hunger, and you might now be convinced that this idea is not merely a lark. If this is something you really want to do, follow the path. You'll never know until you try. You must! Do it!

If you're a civilian in college or a graduate, then visit a Navy, Marine, or Coast Guard recruiter. If you're already in the military, your commanding officer or his staff will tell you whether a path exists. If a program is open, submit a formal application.

Either way, your application could eventually wind up in front of the appropriate selection board of the Navy, the Marine Corps, or the Coast Guard. That will be a crucial moment. That step determines if you will be offered the opportunity to be considered.

When you decide to jump into the fray and to seek assignment to Naval Flight training, you might think you have come into contact with a host of personnel digging and clawing for every flaw you have. No! No! Wait up! Actually, they are looking for reasons NOT to reject you. They're anxious to find candidates. But they don't want merely a warm body. They want a qualified candidate.

Your stars must be in their proper places. Your mindset must be aimed in the right direction. You must demonstrate an elevated aptitude to flying. You must verify you have made high-quality use of your time in school. If you are applying as a current member of the armed forces, you must have had an exemplary record. And further, by the Grace of God, you must have been born with the physical, intellectual, psychological, and emotional attributes required for success in Naval Flight Training. Could you reasonably expect that anyone will be selected to sit in the cockpit of a multi-million dollar airplane with less than the proper foundation and ability? Finally, findings are submitted to the appropriate military service.

Once the flight selection board receives your application, they go to work behind the scenes. Military and civilian officials check your background. Drug problems? Law enforcement profile? College grades and attendance records? They'll probably talk to neighbors in your home town. Nothing is off the table. You'll never know unless some friend greets you one day with, "Hi. Somebody from the government dropped by last week and asked some questions about you."

If you pass that filter, you are invited to an appropriate Naval base and guided around the white, etherized walls of a medical center where it seems that pints and quarts of medicines line the bulkheads in little vials. Some applicants probably have had thoughts cross their minds as to how they keep all these meds straight. Corpsmen are professional.

Eye tests may be an obstacle to some candidates. Any thing less than 20/20, two tests in succession, has sent many a hopeful to the lockers in spite of any amount of pleading. The medical people dilate your eyes and look into them, observing biological structures very deep inside this complex cavity. The retina must be squeaky clean. Where did they find all these tests? You didn't know all that existed.

If you're still in the running, the shrinks get their hands on you. When the technician points to a strange looking picture during the Rorschach test, say "monster" if it looks like a monster. No second-guessing or joking around.

The doctors or technicians might put you on a couch or a chair in a dark room and draw out answers from your deep thoughts. Don't worry. They won't be broadcasting your responses as food for a joke when they eat lunch at noon at the officers club.

If you are deemed to be a well-rounded individual with no strange dark mysteries in your past, the dental officer will certainly want you to visit him or her. At different times in history, the Navy has had different standards. Sometimes applicants with a couple of untreated cavities would temporarily disqualify an applicant. At other times they might allow an applicant to be missing a few teeth. A hint is offered here: Go to a civilian dental check before and get a checkup and treatment before you go.

So you're still in the running? Then head down the passageway to the MDs and their corpsmen. They'll check your pulse and blood pressure and find reasons to look into the many orifices of your body. The Marines have a slogan, "The Marines are Looking for a Few Good Men and Women." It's apropos for the Navy too, and especially for flight student applicants.

Check out in top shape? Good. That's what the Navy wants. You'll go home or back to your unit. Celebrate a little. A little? Not too fiercely. You don't need a peck of festive trouble to spoil your good fortune.

In good time, your orders arrive. You are in, and for up to a couple years thereafter, you will face many successes and hopefully few rejections as a student. Physical and emotional strain continues until graduation day. Hope be it that you will continue success in your training. This might seem a bit impersonal. It has to be. The Navy must put emotions aside. Our country depends upon it.

Avengers of Pearl Harbor

Marine Colonel and Naval Aviator George Dodenhoff has lived in Tampa since his retirement in the 1970s. He plays golf and meets friends, mostly retired military aviators, for morning coffee breaks at a local restaurant a couple times a week. It is one of those thousands of "breakfast clubs" that inhabit restaurants around the country. They deserve it. His wife, Priscilla, plays a great hand of bridge while the colonel directs his pimpled golf ball toward all those little four and a quarter inch wide holes.

I was interviewing this 91-year-old super gentleman for this book. Remembering the absolute respect for seniors that I had learned both at home as a kid and reinforced by my Marine Corps drill instructors, I addressed him as "Colonel" and "Sir" a couple of times. He stopped me. "Bob, I'm Dode (Dodee). From here out, you call me Dode. I'll call you Bob. Besides, we're both Marines and aviators – you can't do better than that."

Dode had a career that few can match. He loved talking about flying and his over thirty years in the Corps. I believe he would have talked for days and I could have listened to every word about his time in WWII, the Congo, Korea, and Vietnam. Then I had to tell him that for this book, I was specifically looking for stories about his days in Naval Aviation Flight Training.

He thought for a few seconds. "OK. One comes to mind," he said. "How about the story of how I joined up." What else could be better? He started his story:

> *In June of 1942, six months after the Japanese bombed Pearl Harbor, I was eighteen, living in Brooklyn, NY, and waiting for my draft call. I was attending Brooklyn Polytechnic Institute during the daytime and working the night shift in a defense plant making electronic devices for Navy aircraft.*
>
> *A good friend and co-worker of the same age suggested that we should join the Navy, be pilots, and help get this war ended soon. I thought that was a good idea so we tracked down a recruiter. My buddy failed his*

physical but I passed all my tests and they told me I'd be notified what to do next.

Time went by. I kept working. I kept going to school. I began to think they had forgotten me. I went back to see the recruiter and he told me to hang on. "Do just what you're doing. Stay in school, keep working, and wait till you hear from the Navy."

Then one day I got the letter. They told me to report to a certain Brooklyn radio station at 7:00 pm, Monday, December 7, 1942. That was exactly one year after the Japs attacked Pearl Harbor. I thought that was a strange request, but if that's what they wanted, I'd do it.

When I got to the radio station, they put me in a room with some other young men. They told us we were about to be sworn in to the U.S. Navy. FINALLY! But why the radio station?

At a few minutes after 8:00 p.m., they crowded us around a microphone and turned on a radio speaker. The speaker blared the voice of the great Fred Waring of the nationally syndicated variety show, "THE FRED WARING RADIO SHOW."

Mr. Waring announced that in one minute a Naval Officer down in Washington would come on the mike and speak to young men standing in radio stations, just like we were doing, all over the country.

And he did. That night, 10,000 young men, all over the nation, were sworn in simultaneously as Naval Aviation Cadets, over the airwaves of his show.

Afterwards, Mr. Waring took the mike and crowned us "The Avengers of Pearl Harbor" as he thanked us for joining the fight to save our country.

"Bob," Dode said to me, "I was never so proud." So am I, Dode. Can this not be a high point of Naval Aviation? A sad note: Dode died on August 6, 2015, before this book was published. It is my hope that when he reads it from the heights of the Pearly Gates, he'll approve.

As I prepared to write this book, I received many stories from former Naval Aviators and students. From these individuals came humorous, nostalgic, tragic and informative accounts. The following stories are in this The Road to Pensacola chapter because they say a lot about what sets the stage about decisions we might have made at that crucial time of our lives.

Visit to the Crash Site

The author of this book, known as Bobby at the time it took place during WWII, submits this story. Bobby lived on a dairy farm near Augusta, Georgia and even nearer to Camp Gordon and Bush Field. Camp Gordon was a training base for U.S. Army infantry troops. These brave soldiers would get their final preparation before being sent to England to fight. Bush Field was an Army Air Corps basic flight training site for student pilots training in Army BT-13s.

Everyone seemed to know our country was headed for war. Some tyrant named Adolf Hitler was running rampant over the countries in Europe and he kept saying he would rule the world. Some parents tried to shelter kids from the meaning of war. Most kids were as aware as they could be at their young age.

> *I was eight years old in 1941 and obsessed with airplanes. Why not? They were from Bush Field and were Army Air Corps BT-13 trainers. They were constantly diving and looping all over the sky. I wished so much I could just reach up and grab one.*
>
> *I did touch an airplane one summer when a BT-13 crash-landed in my granddaddy's corn field. The Army posted a 24-hour guard and no one was allowed nearby.*
>
> *One day my granddaddy and I were in his truck headed to Mr. Johnson's store for a 3-Centa, that soft drink much like a Coca Cola, but it only cost three cents. Smart, isn't it, and we could split it. We had heard that a plane had crashed on our farm recently and my granddaddy was trying to find it for me.*

"There it is, Bobby," he said. We slowed as we passed by some corn shocks and a plane nearby. It looked much bigger at a hundred feet away than it did in the sky. We stopped and walked toward the plane. We knew we weren't supposed to get close to the planes so we stopped about at about ten yards away.

Still, we didn't see a guard and I asked my granddaddy if we could go a little closer and touch it. I'd have something really exciting to tell my buddies in school. We got out and walked closer. Feeling a bit adventurous, my granddaddy lifted me up onto the wing. What a thrill! The joy lasted for only a minute when a guard raced up, arms waving, screaming for us to move back.

We talked to the guard for a few minutes and my granddaddy explained that this field was his cornfield and we just wanted to see a plane up close. The guard relaxed. He even lifted me into the cockpit and I began to punch dials and switches and buttons on the dashboard of the plane. He asked if I wanted to be a pilot when I grew up. I answered a heavy yes sir and he told me I'd have to study hard in school.

I was taken with that experience. I knew that one day I'd be up in the sky like those men at Bush Field. I promised myself I'd study hard so I wouldn't crash. Once when I was ten or so, I was brave enough to tell my mama that I wished the war would last long enough so I could join the Air Corps. I thought my mama was going to paddle me, but she calmed down with just a tongue-lashing. Mama had a brother in Europe as a bazooka man in the Old Hickory Division (30th) and another based in Recife, Brazil as a torpedo man aboard a sub-hunter. My family had a heavy investment in this war, but not an unnecessary one.

First Plane Ride

A former Marine aviator, Robert, contributed this story of how graciously his Aunt Grace treated him. Not only did he tell this story to me by phone but he also sent me a copy of original hand-written notes his Aunt Grace made shortly after this happened. Recently, when she knew she'd soon pass away, she sent him her notes. Robert said that as he read them, he remembered that day, but I doubt that it was as vivid as was her description. Robert said that what she had wedged in his mind for several years afterward might have had a small part in his becoming a Marine Corps aviator.

My name is Robert. Quite often during World War II when I was about ten, my mama and I would drive to my grandmother's farm on Mayfair Road near Sumter, S.C. to visit all of her brothers and sisters. Twenty-year-old Aunt Grace would be there. She was my favorite person in the whole world. She said I was her favorite nephew. Later I realized I was her only nephew at the time.

Aunt Grace knew how much I loved airplanes. When I would visit their house, we would often walk out in the pasture and look up in the sky. We'd have no trouble finding them. Sumter Army Air Field was about five or six miles away and airplanes flew constantly to and from that field. As some aviators are prone to do when they are out on a training mission, they fly low over the farms and villages, looking for a target to buzz. When one of these planes would fly over us, I would jump up and down and run and scream.

One day in 1943 when Mama and I went to visit, Aunt Grace grabbed a baseball and a bat and told me we should go out to the pasture and play. Gosh, I liked baseball almost as much as watching planes fly. I was out of the door in no time.

Aunt Grace told me to keep looking up into the sky and if I saw any airplanes, I should tell her if that was my favorite. Of course, every plane I saw was the one I liked best. In a few minutes, I saw a plane that seemed to be getting closer. I yelled to Aunt Grace as we stood and watched it circle around us several times.

"Look, Robert," Aunt Grace said. "It's getting really close now."

"Looks like it might come down where we are."

The plane seemed to go away, but then it turned to come back toward us. Every second, it appeared to drop closer and closer to the ground and get bigger.

"Aunt Grace! He's coming down. I'll run over and see him."

"No, you stay right here until we see what it's going to do."

The pasture had some smooth spots and on one of them was precisely where the plane touched down and came to a halt.

"Look. Now, can I run over, Aunt Grace?"

"No, not while his fan is turning."

"Now it's quit, Robert. Hold my hand and let's run and see the pilots." Not only did I see the pilots but the pilot in the back seat got out of the cockpit, picked me up, and sat me in the seat.

By now the family was running from the house. My mama saw me sitting in the cockpit and had one of those conniptions mothers have when they think their kid is in peril.

Aunt Grace whispered something to the pilot in the plane. He started the engine and began taxiing up and down the pasture with me still in the back seat. My mama yelled at the top of her lungs but with the engine running, but I couldn't tell what she was saying. In a few minutes the pilot shut down the plane and lifted me out. "Can I go flying?" I kept asking.

The pilot who had gotten out reached in the plane and pulled out something wrapped in a napkin. He opened it and held it up. It was a set of Army Air Corps pilot wings. He pinned them on my shirt. Wow!

The last thing I told the pilots just before they started the engine to leave was, "I won't ever take my wings off!"

The whole family, including a few farm hands, had gathered to get a first hand look. We all watched as long as we could see a dot in the sky. I waved until I could see the little dot no longer, but I probably thought I could. For a long time after that, on each visit to Grandmother's I'd stand and wait to see if the plane would come back.

Although the truth was never mentioned, at least in front of the kids, the family must have been sure that Aunt Grace found some way to arrange that L-19 fly in. Robert grew up, finished college, was accepted into Naval Flight Training, and earned his Navy Wings of Gold. He states that even today he will touch his set of wings and the memories travel back to that day in 1943.

Gotta Go

Howard had been in college for about a year and a half, just a few months before the Japanese attacked Pearl Harbor. When they bombed us, he began to ache to get into the war as a Naval Aviator. I interviewed Howard for this story over the telephone and he got my assurance that I would not mention his real name or even his home state.

I was about to go nuts. School was driving me up the wall. I'd been there since before the Nips surprise-attacked Pearl Harbor. As soon as that happened, I wanted to get into the Navy and take this war back to them.

The Navy recruiter looked at my school records and said I needed more credits. He told me to stay another quarter or so and the Navy might be able to wave a some credits of the two-year requirement.

I had this idea. If I'd get a job in the dean's office I might be able to see how the records are kept. Maybe I could fix my records and maybe even add some credits. After a couple of interviews I got the job. I was the only boy in the office but that didn't matter.

There was this cute little girl working in the office. I had my eye on her anyway. She was a smart student and one of her jobs was to transfer grades from instructor grade sheets to permanent records. She was a real stickler about entering grades correctly and keeping them locked up and secret. The only person senior to her regarding grades was the dean.

After a while I noticed she kind of had her eye on me. We had a few dates and I kept asking her if I could do more important jobs. I knew that if I could, this might put me in position to fix my grades in line with the Navy's requirements. She started dropping her guard.

For a while, she would never leave the grade safe unlocked when she had to leave. Then one day she told me I was doing a good job and she thought I was trustworthy and she could leave me in charge and take some of the load off her. I promised to do a good job. I knew I had to be careful.

My chance came. She asked me if I could increase my hours on the job. Her grandmother was really sick and she would be out for a few days. I had it all to myself.

I played it cool the first day, except for a few practice runs as I opened the books and thumbed through, casing the sections that I would get back to later. I rehearsed exactly how to do it. This would be a one-shot undertaking. I could not take the chance of making a mistake and doing it again.

I made some notes reminding myself of the two courses I had failed earlier. I realized they could be easily changed to passing as long as the professors didn't review the permanent records.

I waited until lunch and told the other two clerks to go on without me. I said I had a couple of entries that needed to be made. Perfect! It only took a few minutes. I even made it to lunch before the dining hall closed.

The end of the quarter passed and we recorded grades. I made sure I recorded all my grades. I changed one of them. It was time to see the Navy recruiter. When he sent the request for my transcript, I even filled out the transcript and mailed it back.

A few weeks later, the recruiter informed me to come in to get my paperwork so I could get my orders to Pensacola. He didn't even notice that my former failing grades had been changed. If he did I guess he just needed a warm body with adequate grades.

I spent my war years in the Pacific aboard a carrier. When I returned home I returned to college but not the same college. I got a degree, but not in the same state.

Christmas Dinner

For a military person transferring from one duty station to another, mainly with a family, and more especially at Christmas, life can be traumatic. As the author of this book, I offer this example of the beautiful relationship between NAS Pensacola and the City of Pensacola:

One of the most outstanding attributes of Pensacola is the relationship between citizens and the Navy. The city is definitely one of those Cities of Brotherly Love.

I was a Marine infantry officer in 1959, recently home from an overseas stint in the Middle East, specifically Lebanon. Before deploying to Lebanon, I had been accepted for flight school, but the Middle East situation tightened, I was "frozen" and could not transfer.

Nine months later when my unit returned, the Marine Corps reactivated my flight training orders ordering me to report by Christmas. My wife and I agreed that it was the absolute worst time of the year to check into the main gate of a new duty station.

We arrived on Friday; Christmas Day. We had driven from Camp Lejeune, North Carolina as my orders demanded, to arrive not later than 25 December, in its use of legal-sounding phrases, including the word "haste". We found a vacancy in a motel in Warrington, checked in about 1700 and unloaded our vehicle.

I drove across the bridge to NAS to check in so as to save a day of very precious leave time. The Marine sentry at the kiosk directed me to the check-in office. The Officer of the Deck signed me in and instructed me to report back on Monday at 0800.

My wife and I looked forward to Christmas dinner. We had been married for almost two years, but hadn't spent a single Christmas together. But we'd have Christmas dinner tonight! The motel clerk told us that not many places were open that night. He suggested we drive along Warrington Road or Navy Boulevard. It was almost 1930 as we cruised slowly down the darkened and almost empty streets. We saw no lights in any restaurant.

Finally, my wife spotted a café with lights and people inside. We parked the car and walked to the door. A handwritten note on the door glared at us with, "Closed for Family Dinner."

We turned to walk away when a gray-haired lady rushed to open the door and said, "We saw your uniform and figured you had just gotten into town. Come in. Have dinner with us."

That night turned out to be one of the most enjoyable holiday meals we ever enjoyed. We wrote to the editor of the Pensacola News-Journal. We went back to eat many times. When I graduated, we, including our new little daughter, ate our farewell meal in our beloved and favorite restaurant in our much-beloved Pensacola.

The restaurant is no longer open, but the ground on which it sat is most definitely hallowed in our memories. Good and unselfish people owned that business and shared with us. During the year and a half we were in

Pensacola we saw many other examples of the superb relationship between Pensacola and the Navy.

Carrier Qualifications for an E-5

In the Pacific area in 1944, Tom was a petty officer E-5 in VPB-20 at Leyte Bay, Philippines. To his joy, and after a lot of dedicated effort, he had been selected for pilot training. It didn't work out very well, as you'll read in his story:

In 1944, I was an E-5 in VPB-20, stationed in Leyte Bay in the Philippines. Unfortunately my goal of wanting to fly Navy as an enlisted man was diminishing because the war was winding down. I still had carrier qualification training to go to reach that goal, but I kept realizing that I was still an E-5 and had a long way to go. Most of the full status aviators were officers.

In the military way, everything is done by seniority. My slot in the lineup for carrier flying finally came up and dictated that I was to ship over to the USS Saipan, CVL 48.

I finally arrived and waited in the enlisted berthing in the bowels of the Saipan for my turn to come up so I could continue learning to fly. I realized that my seniority put me nearly, if not, last. Then one day as we approached sundown, I heard my name announced on the PA system to report to the flight deck and fly my six daylight carrier landings. This was it!

Excitedly, I made my way to the flight deck to man my aircraft. The quickest way to get to the flight deck was through Officer's Country. A Marine sentry was on duty at the entrance to O-country and he stopped me cold. He said I was enlisted and had no authority to be in O-country. That ever-dutiful Marine refused my pleas no matter what assurances, exchanges, or pleasantries I gave.

Finally the Air Boss sent a Navy LTJG from the bridge to find why I had not reported. The officer assured the

Marine sentry all was well. I proceeded pronto to the flight deck and quickly was able to make six successful carrier landings, even though the last one was in what the Air Boss said was done in "near-light conditions."

Unfortunately, that was my last flight. Several of us "oldtimers" were given orders back to the States. Navy programs sanctioning enlisted pilots were being terminated.

CHAPTER 8

Preflight

They were fledglings; new men (and women as of February 1974) on the block; mere hatchlings and they acted the part. It was of little importance whether they were just out of two years of college and the rawest of Naval Air Cadets or Aviation Officer Candidates. They might have been Navy LTJG Naval Flight Officers with sea duty experience. They could have been USMC junior officers with a year or two of platoon leader experience. During one period, the Marine Corps selected some O-3s (captains) for flight training. That was just before Vietnam and a knowledge of infantry maneuvers would help. No matter, because this was a new day; a new dawning. This was to be a different war, whether on the ground or in the air.

Seniority or not, their nerves told on them. They were all students. Some of the commissioned students jumped ahead of the normal Preflight syllabus by about six weeks. The rest were diverted and handed off to drill instructors for a couple months to have the military ways pounded into their souls, hearts, and bodies.

Black Tuesday

Mike was an Aviation Officer Candidate in 1968. His drill instructor, unnamed for this story but remembered for life, didn't want even one of his candidates going astray. You might remember, "Do not stray. It's the Navy way." Neither did he want his charges to do anything off-track to anyone else. The sergeant had a reputation of being "as tough as nails."

Black Tuesday was known as that horrible day when Aviation Officer Candidates "bared all." A very detailed personnel and equipment inspection was paired with three of Preflight's most stressful training events; the Dilbert Dunker, the Confidence Course, and the Altitude Chamber. Black Tuesday was appropriately named because few candidates passed personnel inspection.

Mike didn't pass personnel inspection that day and after this many years he said he is finally able to talk about it. He hopes his DI has retired to some desert island and has no wish to even think about any of his former wards.

Mike's story:

> *His name was "Sir", but under my breath I preferred calling him "Nutcase" or worse. But out loud, I called that Marine DI "Sir" or "Sergeant". The way I saw it was I knew he hated AOCs, so why should I like him? He always said he didn't give a rat's ass if we liked him or not because "this is not a popularity contest." It was strange how it seemed to me that he was a different person when he was talking with the other DIs. They were always joking, laughing, and having fun. And when a staff Navy or Marine officer approached him, he was fruity tuttie Mr. Nice.*
>
> *I shared an upstairs room (I should say "topside quarters" in case Nutcase reads this) with a group of other AOCs. Today was Black Tuesday, the toughest inspection we would have in all of Preflight. We finished our Dilbert Dunker event at about 1130 hours, ate chow, and headed for our quarters to prepare for personal inspection. I had laid out all of my equipment and clothing on my bunk the night before and slept on the deck beside my bed, uh rack, sir. That little hint came from some*

former enlisted personnel who were now Aviation Officer Candidates. I had to go with experience.

My roommates and I heard him thundering down the passageway toward our quarters. We snapped to attention when the door, excuse me, the hatch, opened. He looked us over with a glare that I still recall and have nightmares of Candidate Monroe's wallet. "Sir" began to fiddle through it. He pulled out an old picture from his wallet of two elderly gentlemen with white hair and long beards, with their arms over each other's shoulder.

"Who the hell are these people?" he roared at Candidate Monroe.

"Sir, that is my grandfather and one of his cowboy buddies."

"Don't look like no Gene Autrys to me."

The rest of us started laughing, but tried to hold our composure. But we lost it.

He turned toward me. "What the hell is so funny?"

"Nothing, sir."

"Then why are you laughing?"

"No reason, sir."

Still looking at me, he asked, "Where you from, Candidate?"

"Fresno, sir."

"Where?"

"Fresno, sir."

"Oh, up north in California? I thought they called it FREEZNO up there," he said pronouncing it very heavily FREEZNO.

"No, sir. Fresno."

"I know where that is. So the candidate is from FREEZNO. I know where that is. You don't think I know where it is? It had the first real garbage landfill in the United States. You know that?"

"Yes, sir. I knew that."

"Then how come you didn't bring it up?"

"Didn't think it mattered, sir."

"Everything matters about FREEZNO, Candidate."

"Thank you, sir."

"They got grapes there, don't they?"

"Yes, sir."

"Last time I was there I ate a handful of your grapes. Had the runs for a week."

My roommates broke into heavy laughter. I cringed. They couldn't stop. That set him off. He grabbed me by the shoulders and started pushing me toward our door. Suddenly I realized I was headed out into the passageway, probably for a PT session. It should be called SPT because when Nutcase did it, it definitely was Sadistic Physical Training.

The sergeant kept pushing and shoving me as I tried to keep my balance. He started spinning me around. In the shuffle, my shoe soles scraped across the top of his beautifully spit-shined shoes. He looked down.

"Look what you did, Candidate." I looked but refused to believe it. A huge dull brown scrape ran right across the top of that left shoe. It would never be the same. I wanted the deck to open up and swallow me, but it wouldn't. My fellow candidates looked down and started laughing again. They couldn't shut up. As hard as I tried to keep my self-control, I joined in.

The infuriated sergeant grabbed me and rammed me hard against my locker. Where he was probably faking before, he seemed to be serious this time. My locker moved, jostling free a large heavy oscillating fan that had been sitting on the top of the locker. It rocked it so hard it fell off and landed right on top of Nutcase.

Now the sergeant went ballistic. He opened my locker door, shoved me inside, slammed the door shut and single-handedly began walking my locker across the room toward an open window, all the while telling me that in one minute I would be getting my first dose of Navy flight time from an altitude of two decks.

I thank God he stopped short of throwing the locker out of the window. Instead he pulled me out of my locker and sent me to the passageway for PT. I guess

I accomplished PT satisfactorily. I don't really know because my memory began to get hazy.

I came to as he was telling me he'd forget this whole episode if I'd get my people at home to send some FREEZNO grapes. I agreed and wrote my parents to send some to Nutcase but didn't tell them why.

Several days later I was called to the duty office. A group of DIs, including Nutcase, huddled around an open box on a desk. The box held a lug of the prettiest grapes Fresno had to offer. My mom and dad had airfreighted a 25-pound iced down lug of Class A grapes. Nutcase didn't even offer me a single one of those grapes, but I did get a thank you from the DIs, including Nutcase himself.

Several months later I was in Basic Training at Whiting Field, reading in the Whiting BOQ lounge. A couple of AOCs were playing cards at a table when I heard their conversation shift to a story about a guy once in Preflight getting shoved into a wall locker and thrown out of the second story. "Thrown out? Second story?" I thought. Kind of like the old game, Gossip, where a story changes. The AOCs were exaggerating a bit, but I remained quiet.

Recently, my wife and I were visiting friends in Pensacola. Channel Three TV was on and a young reporter was airing some old Navy footage with a retired Marine staff DI, recalling old days. The former DI stated that once he had an AOC in his Preflight platoon from whom he requested that grapes be sent from FREEZNO (he emphasized FREEZNO). He jokingly wished that AOC could know he was appreciative.

From this bolt out of the blue, I realized I had become a legend in my own time, actually in double-time. Thanks, Nutcase. I love you, too.

The Neat Freak

Frank was a Naval Aviation Candidate in 1967. His Preflight Class had the typical count of 32 NavCads.

Statistically, ten to a dozen would not make it out of Preflight into Primary. Illness would take some. Others would have trouble with physical requirements, academics, or discipline. Still others would decide their mama was right, that they weren't cut out for this, and "drop on request." About twenty would go to Primary Flight.

The war in Vietnam was raging to a higher level, requiring the need for more of all types of pilots, both helicopter and fixed wing. The NavCad billeting area grew crowded. Maintenance installed six feet high dividers with curtains so that two NavCads were enclosed within a small space about 12 feet by 12 feet. Inside each cubicle were two very small single beds, two medium-sized clothes lockers and two dressing mirrors. The shower and head were down the hall, midway of all the cubicles.

Frank remembers a neat freak student: "Jerry".

> *Jerry was a pain to put up with. If he were to drop a smidgeon of food in his lap at chow, he'd find some way to get a few minutes to go change. He did it once by faking a gagging attack. If someone dropped a cracker in the squadron recreation hall, he'd gripe until it was cleaned up, or even clean it up himself, demeaning it as he worked.*
>
> *No NavCad wanted to bunk with Jerry. As I might have guessed my fate, Billeting assigned me as his roommate. I tried to be friends. That's what I thought I should do.*
>
> *I rarely saw Jerry in the shower stalls. When I did see him he was wrapped up in a huge beach towel and doing nothing but shaving.*
>
> *I began to wonder if he actually ever took a shower. But he certainly looked clean. I mentioned this to some of my friends and found that they had been curious too. Several of us began to wonder if he might even be a girl, but then decided that Navy corpsmen, during all those physicals, couldn't miss anything as evident as that.*
>
> *One morning at roll call, one of my friends, Skip, told me he had gotten up about 3:00 the night before for a*

head call and he actually saw Jerry finishing his shower. Skip hid in a stall as Jerry dried off, wrapped himself in a towel, and returned to his cubicle. Skip assured us that Jerry was not a female, even though he had to squint really hard to make a distinct decision.

I felt sorry for Jerry. I was in the majority in thinking that he was not in the right place at this time of his life. Anyway, I had my own future to consider, so a couple of weeks later I went in and had a discussion with Billeting. They reassigned me by giving some cock and bull story to Jerry. That left the guy living solo. That was probably what he wanted anyway.

He had no friends. We doubted that he'd ever had a date. He lived a lonely life. Strange substances mysteriously began to collect under or near his bunk: a rock with a kind of sunset scene, a leaf with a perfect circle decayed in the center, a leather belt chewed almost in half. That scenario could drive a psychiatrist to a Section Eight himself.

Out of a conversation with Skip, we talked to some of the other members of our squad. We came to the conclusion that we'd go downtown to the bar area the next weekend and try to find a sweet young thing for Jerry.

The lounge was packed, but then we saw her. There she was, sitting on a stool at the bar. Skip and I walked over and stood beside her. An immediate smile on her face told us we might have picked the right person. We learned that her name was Annie. After buying Annie a couple and assuring her we were not ogres, we explained our mission. She was willing. Finally we would get Jerry on the right track.

It took most of the following week to convince Jerry to go with us. After we offered to buy him a nice juicy steak dinner at Steaks-a-Lot on Saturday, his favorite place, he consented. We arranged with Annie to meet us the next weekend at the bar.

Saturday night at 9:00, after our steak dinner, Skip, Jerry and I walked into the bar and sat at an empty table. A few our squad followed us in, and so as not to create a theatrical atmosphere, sat hidden at a back table. Annie was seated on her stool and rushed over to our table when she saw us.

The three of us stood. When Skip introduced Jerry to Annie, his eyes twinkled just a bit. After about 10 minutes, and half that many beers, Annie and Jerry were absorbed in heavy conversation and Skip hardly noticed as we quietly left the scene.

At the conclusion of our Preflight class of NavCads, we splintered and shuffled to our appropriate Primary training fields. Annie and Jerry were still together and he no longer gave a hoot if a spoonful of soup fell on his shirt.

Dilbert Dunker

The author gets to put in his few cents worth in this story. Even before I was selected for flight school, I had heard rumors that down near a city called Pensacola there was this nearby Navy base that possessed a bizarre apparatus into which frightened humans were stuffed and ordered to wait for some crazed human on a tower to push a button and condemn them to horrible deep sea fates. It was called the Dilbert Dunker. I was a Marine and I was sure it was only a dirty rumor started by some Navy guy who hated Marines. I soon discovered the Dilbert Dunker was real. After training in the Dunker, I saw what it taught and understood the reasons we were required to undergo that training.

In 1959, I was a Marine with three years in the Infantry. I got the call to "Fly Navy" as a Marine. This story might be told by saying, "Twice during flight school, all within a couple of hours, I rode the Dunker down the long, long tracks. Everyone had to do it. It was required." But there is more to this tale.

My wife had attended a wives orientation to Naval Aviation Flight Training a few weeks before my Dunker training date. After her orientation, she kept teasing me about the Dunker. She asked if she could take my place and ride it down for me. As the day grew closer, I was getting a bit twitchy and began to consider it.

Most of us remember Dilbert Dunker day. We were in Preflight and had not even seen the inside of a plane. One by one they called our names and we lined up at a station leading to the ladder to that ride that would come soon. Few things before had ever caused me such apprehension — not even climbing down cargo nets from APAs into landing craft at three in the morning in raging seas.

We'd each take one run in the Dunker; down that monster that someone once said, actually smiles as you get in it. As soon as Williams, the last man in the class, was down the tracks Bingham would be at the top ready to go his second and final turn.

However, one of us, a Marine lieutenant named Gale, received permission to take his two turns in a row, so he moved to the head of the line. He was being married on Saturday morning and had reservations that afternoon for a flight to Missouri. He promised that his new wife, Janet, one of the nicest people in the world, would have him back Monday morning, if he somehow survived his all day Sunday honeymoon.

Gale received loud cheers as he came down the Dunker on turn one and surfaced. Then he scampered back up the ladder for turn two. The Dunker area thundered with shouting and thunderous applause as he surfaced from turn two. Gale was finished. Parrying catcalls chiding him for his decision to take a bride, he picked up his gear. The catcalls turned to applause, even from the staff officers and attendants. A display of camaraderie!

Gale actually was back on Monday in time for his first class that morning. He received more of the usual nagging that new bridegrooms endure for a while after the wedding.

Time passed and he soon had completely shed traces of his bachelor rank. He and I and our wives hit it off. We became inseparable for the remainder of our time in Flight School. Upon graduation we both received orders to sister squadrons at El Toro. The four of us continued our friendship of playing bridge, going to dinner, oohing and aahing at our newborns, attending christenings, birthdays and grand visits to Disneyland.

Scavenger

Some marginal students slip through the fight school selection filter. Heaven knows how. Gus was in Preflight in 1972 with a marginal Naval Cadet named Alex. Two months or so into the program, Gus was sure he himself had stacked pretty well in graduating from Preflight and heading to Primary, but he had serious doubts about this kid, his roommate, Alex. Alex was klutzy in every direction. As in every class from the beginning of time, little groups of NavCads sit around and talk about the others. Gus sends this story:

No one expected our fellow NavCad friend, my roommate, Alex, to make it through Preflight, much less solo, and Heaven forbid, get his wings. We used to tittle-tattle about him and discuss how we thought he was wasting his time and the Navy's time and money. We thought he felt that way too. At one point we believed he had given up inside, but had refused to accept it.

Alex topped the scales at well over 200 pounds and Medical put him on a weight reduction program. On one of his weekly visits to Sick Bay, his flight surgeon told him he had to work harder; that he hadn't lost quickly enough. In a joking moment the corpsman told him if he got through the program, they'd probably only let him fly transports; that no other class aircraft could lift him. Alex accepted comments like this without a smile or a frown. He probably had become accustomed to put-downs for years.

Then we learned that Alex's father had been an Air Force F-86 pilot. In 1953, his dad had been shot down by a SAM over North Korea and never returned nor was ever heard from. We thought maybe he told us about that so maybe we'd lighten up on him. If so, it worked. We quit any form of mockery.

Then came Survival Week. Those swampy woods, including the slow moving stream, down near that Air Force Base we learned later was Eglin, were just plain scary. Some of the guys in our class had never seen a snake but they certainly had their fill during that week. "Food?" they would say. "How are we going to find food down here?"

One guy, Davy, even caught some snakes and made a pot of snake soup. A few of the NavCads tried some. They offered some to Alex. He looked down in that cup, saw the snake parts floating, and immediately devoured it. What? Alex?

The last two days we had our March to Freedom. At 1500 on Wednesday, we were divided into several teams of four. Each man received half of a parachute, a marching compass, and a piece of a map with a route outlined that covered 20 miles through the swamp and a river and briar patches and yes, snakes.

We were told that squads of "enemy patrols" would look for us. If we were caught we'd be taken back to the starting point, issued a new map for a different route and started again.

When we organized for our march at darkness, we asked Alex to be our guide, our leader. The guy accepted. He looked at the compass and the map then looked up. The sky showed a half moon. Alex told us we should do OK but if we had a full moon we'd do better.

Our leader, Alex, went into action. He gave us our orders. We'd only move in darkness and fast. At the first indication of the enemy we'd drop flat in our tracks. We would catch our food during our marching, hold it, and when we found cover during the day we would eat it; cooked or raw.

It worked. Alex kept up with our progress and the first day he determined that we had covered over 12 miles. Now we had only eight to go with a full night to travel, if weren't caught. We could cross the finish line in a bit less than eight miles.

We succeeded. The enemy patrols did not spot us. As we debriefed, Alex commented on his dad and wondered how long he might have evaded and escaped.

After that, all of us knew that if Alex's father was anything like Alex, the man was still escaping and evading in North Korea.

Dad's Revenge

Elliot and Hal were roommates in college. Both names are aliases because for whatever reason, Elliot desired anonymity. Hal was tired of college and decided to perk up his life and fly Navy. I would not be surprised if there were hundreds of similar occurrences during the Vietnam Era.

Elliot wanted no part of flight school. He served his time in Vietnam in 1968 in the Naval Construction Battalion (Sea Bees). Read Elliot's story:

My college roommate, I'll call him Hal, at the University of North Carolina, grew tired by the day of studying. Then one day he hit me with an idea. He asked me to get in the boat with him. He suggested we quit school, join the Navy, and fly jets. For a couple weeks he tried to talk me into joining him and even found a nearby Navy recruiter who would sign us up as a NavCad. "Besides," he would say, "after the war is over we can come back on the GI Bill and get a free education."

I didn't want any part of flying, and especially if somebody was shooting back. So Hal set out to do it on his own. One weekend he left without letting me know.

That was OK. He always did that. On Sunday he popped in to say that he'd be leaving in a few days. He signed up to go to Raleigh and see if he qualified to Fly Navy. If he passed, he'd be heading to Pensacola for blue skies, girls, and plenty of his own money to spend. He wouldn't have to argue with his dad any more about all his spending.

He promised me in about a year and a half when he finished flight school, he'd drop back to visit me before he went out to sea. He'd want me to see how his shiny wings would look on his very own ensign uniform. "I'll probably have a car too," he said. "Be sure you find us a couple of sweeties and we'll take them over to the old airport and park till the wee hours."

I asked him what his family thought. He hadn't told them yet but he would before he left for Pensacola. We shook hands and he left.

Six weeks went by. I heard nothing. Not even a card. I began to worry. I didn't know how long it would be before he would start flying so I wondered if maybe he'd crashed. Then some of my friends told me that in flight school a student doesn't get near an airplane for at least four or five months.

Then one day I was sitting at my study table in my room when the door flew open. There was Hal, no uniform, just civilian clothes. I asked him what he was doing.

"I can't believe that place," he told me. "I thought I had it tough here." He told me they were out nights running through woods where snakes and wild pigs live. He said they jumped through obstacle courses and got back at 01:00 in the morning with an hour or two of studying to do. Then on Saturday mornings there were inspections. He thought they were supposed to fly.

"And every morning except Sunday was a real bummer," he said, "We'd have to get up at five and scrub everything germ free." He told me that even on Sundays some NavCads had to do some kind of dirty duty if they had goofed off during the week."

He looked at me and said, "I ain't having that. I guess I'll suffer when I tell my dad. He'll be pissed, but he always wanted me to get a degree so maybe I can come back."

I didn't hear from Hal for a couple weeks. Then one day he showed up with a couple of suitcases.

"What's up, Buddy?"

"Life sucks," he said. "I went home and my dad got furious. He put me in the car and we drove to Pensacola. Dad said we were going there to re-up as he called it. We had a hard time getting through the front gate, but I showed them my old set of orders. The guard got on the phone and a few minutes later directed us to my old company office. My company commander looked like he was waiting."

"What happened, man?" I said to Hal.

"My dad went in with me. My old company commander told my dad that he would not reinstate me. The CO looked at me and read me the riot act. If I wanted to go to flight school again, I'd have to start all over.

"On the way home I asked my dad if I could get back in school. He told me that I certainly could but he was not paying a dime of it. He said he'd give me a couple hundred dollars to get started but from that day forward it was all my responsibility.

"So I'm here, buddy. Today I'm going to town. I've got to find a job, and I'm going to finish school."

And Hal did just that. Funny what you do when you have to.

CHAPTER 9

Primary Training

Preflight is in our rear view mirror now. We are ascending to the level we have dreamed of for so long. We're ready to break the bonds of earth and soar with the eagles, the Navy eagles. Nevertheless we'll be grounded for a couple more weeks and confined to classrooms. Ground school. We'll learn about the plane we'll fly, the T34-B, its engine and its airframe, its aerodynamic characteristics, its performance and every important and helpful bit of information about it.

We'll visit the bail out trainer and learn to bail out in case of a serious problem without losing body parts or a life. We'll spend a couple of hours in that claustrophobic altitude chamber learning what happens to us at high altitude. This is all good stuff. It might save our lives one day. But anyway you cut it, it's frustrating to gaze out of the classroom windows, across the field and see those planes taxiing, taking off, and landing, and know that even though the time draws closer, it remains in the future. Be that as it may, we now have something to write home about. We plunge into ground school with vigor.

Finally we get our flight gear and our squadron assignments. We walk over to our squadron area for a gear locker assignment. We see that Big Board we've heard so much about. It's really just a blackboard like the ones from grammar school, but our names will be up there tomorrow as we fly our A-1 flight.

A-stage indoctrinates us to survive thousands of feet above the earth in a device that some say, even God wouldn't have intended for man to master. But we do as we progress through landings and takeoffs, stalls, emergencies, and tose long-anticipated spins that so many latter stage students warned us were so terrifying. Not so. After two or three we actually get the hang of it and hunt down new students to tell them how horrible they are.

After about 20 hours of flight time most of us have soloed, except for those who have been dropped or dropped on request. B-Stage tightens up our handle on the aircraft. We hold altitude and airspeed. We learn to perform all sorts of maneuvers much wilder than the spin; the roll, the barrel roll, the loop, the chandelle, full Cuban eight, the half Cuban eight, and its big brother, the Immelmann. In fact some instructors secretly teach us maneuvers banned from the syllabus; for instance, the split S. We won't tell.

We graduate from Primary feeling as if we own the world. We have about 40 hours or so in our logbook. We're ready to hit Basic Stage running. But certainly we'll find a host of new challenges, trials, and disappointments for some. It's the Navy way.

The Gosport

Few of today's aviators have even a slight knowledge of the background of some of the important devices that were invented during the early years of aviation. Back then, so many innovations flooded from the aviation workshops and into the aircraft, that pilots probably said, "Why didn't I think of that?"

Owen, from South Carolina, sends this story:

My days in flight school began in 1942. World War II was barely six months old and anyone who read the paper or listened to the radio could wonder what our outcome would be. Japan and Germany were huge on the horizon; ferocious animals ready to devour the entire world.

My brother, Sam, and I knew we had to do something. He said he'd like to go into the Navy and serve his time on the high seas. I agreed that it sounded pretty good, so we went together to the recruiting station. We signed up and they told us we'd hear in a week or so about when and where to appear for our physicals. They suggested we would not have time to finish our current college term, so enjoy life while we could.

A few days later, we got our reporting dates. We had to tell our mom. She was so scared when we left for Columbia. We promised to be back in a day or two after they looked us over. We knew she was worried. What parent wouldn't be?

We were in Columbia for almost a week. The place was really busy. One day a sailor with a lot of stripes on his arms (I know now he was a chief) asked if we wanted to try for aviation, maybe even for pilot school. We told him why not. He took us to the next building. Sam was rejected quickly because of his eyes and they sent him back to the first building.

I passed, but I'd have to ask my college to send them a copy of my grades. Then I'd hear back from them. Sam was told he could expect to hear in a few days when he had to report.

We went home. We told my mom that Sam would be going in a week. She looked at me. When I didn't answer, she asked if I failed. I told her they would contact me. I felt like I was lying. I knew I'd have to fess up soon, but right then, later was better for me.

Sam left. One day I had been fishing and came in about four in the afternoon. "I opened your mail from the Navy," she said. "Why didn't you tell me you were going to fly?" I hated not telling her. I left for Pasco, Washington in two days.

I hadn't seen more than a dozen airplanes in my life. In fact, most people were still calling them aeroplanes. It seemed like a long time but I finally got through Preflight and into Primary. That Stearman was a hoot. They had a number of N2s that looked awesome to me and I still hadn't gotten off the ground.

Finally, I met my instructor, LT Alexis Ravone, a French exchange pilot. I had a hard time understanding him and he said it was mutual. He told me he wasn't sure he could speak Southern. For a couple of hours, he taught me to preflight the Stearman, how to fuel it, check the oil, and how to do all sorts of plane maintenance. He told me this story about how once he was flying with a student and the engine started roughing out. (I learned that meant running poorly.) He told the student to take the controls and hold the wings level. Then Ravonne climbed out onto the wing, opened a little access door and adjusted something. It worked or either he was trying to scare me.

Then he said it was time for us to see what it was like to fly this plane because we were going for a ride. I started wishing I had signed on with Sam to serve on the ships. We climbed onto the wing and he showed me how to get strapped in. The only adjustment the seat had was to fluff up the thin seat cushion as much as possible.

Soon I was in. I told LT. Ravonne that if I had to sit here, we might as well get started. "Not yet," he said as he picked up a strange looking tube. "You have to get the Gosport hooked up. See, here it is."

"What's that? Way to relieve myself if I have to go?"

"No, it's so I can talk to you. And don't try using it to relieve yourself. The Navy won't take too well with that."

Ravone hooked this inch-wide flexible hose to the ear part of my flight hat. He reached back into his seat and got the mouthpiece and said a few words. The last words were, "Do you hear what I'm saying?"

"Sure do, sir. But how do I answer you back?"

He picked up his mouthpiece and said in a loud voice, "You don't. It only matters what I say and for you to hear it." Then he gave me instructions for the use of hand signals.

He said he was thinking about getting a Gosport and fitting it to his wife. It was about the only way he could get his two cents worth in.

I hated the Gosport. I was always answering without thinking and that caused Ravone to hit me on the head instead of using hand signals. He'd remove the joystick from its slot and whack me on my head. It took a while but I finally got used to it.

Later, after I graduated and was in the Pacific theater, I flew a bomber. My gunner and I used a Gosport but the kind we had was so that we could both talk and listen. It was a lot better but it didn't work too well when both of us tried to talk at the same time.

Idiosyncratic Flight Instructors

Instructors are probably selected more for their cleverness in drawing out the best in a student rather than on their flying skills. Remember the old, old saying, "You can teach monkeys to fly, but they can't make the voice reports." That's not just humorous; that's dead on. This tidbit comes from Dan, the Man:

I was a Marine platoon leader and got the opportunity to go to flight school. During my year and a half there, I met and tried to accommodate every style and make of instructor imaginable. Most hoped that I would succeed and they would go the extra mile when I was having difficulty. Some seemed to be waiting and hoping for orders to Las Vegas or Reno to come in so they could "get out of Dodge."

And always, lurking in the shadows was an instructor who had either had had a bad hair day, a horrible early life, or as a Naval officer, hated Marines, or vice versa.

I, as a Marine, ran across one of those Marine-haters. I will truthfully state I was happy he was my only one. He enjoyed vocally, enthusiastically, and openly, so all in the ready room might hear, placing responsibility on the Navy for lowering the standards for entry into the flight program by selecting lowly me. He gave me my only down flight in my entire flight school career. It was later retracted by a LTCDR flight check instructor who, when he flew my re-check, knew how to bring out the best in a student.

The Ladies' Man

Some students became too daring, even though we didn't know the meaning of the word — much like when we were learning to drive automobiles. But with each phase of training we would become more confident. The secret to our success was to gain confidence without becoming self-satisfied. That was, I assume, the reason for all of the "Complacency Kills" signs displayed around the training areas.

This story submitted by Hubert, a former Naval flight student, doesn't want his last name to appear. It is obvious why:

I was overly proud I hadn't run into any trouble with ground school or flying. In June 1958, sporting over 30 hours of flight time and not a single unsatisfactory flight — not even a negative comment, I began T-34 acrobatic stage.

I spent several nights a week and on the weekends at Limbo Louie's, a little neighborhood bar in Warrington. Some pretty classy females working at NAS also frequented Limbo Louie's. I always had said I loved airplanes, and did, but pretty females were at the top of my list.

I was playing hero to Nancy, one of my favorite classy females, one night about how I'd like to get her aboard one of my solo flights one day. Soon I had opened my mouth once too often. Nancy put me to the test. "Name the time," she demanded.

"OK. Late afternoon on Tuesday, three o'clock launch, I have a solo. I'm supposed to practice acrobatics, but I'll fly over to the practice landing strip just north of Brewton. You drive over and park on the west side behind a bunch of scrub oak trees. Get there about three thirty."

"How will I know you?"

"I'll watch for your car. When it's safe, I'll land and taxi close to where you'll be parked. Then I'll stop the plane. I'll wave a white handkerchief out of the canopy. Then you run to the plane and climb in."

With the plan all set, I took off from Saufley at Tuesday at 3:00. As I flew, I thought and thought and saw no reason the plan wouldn't work. I left the traffic pattern, headed north toward Brewton, and let down to a thousand feet. The only worry I had was that the Crash Crew might discover this scenario, so I decided to circle the field a few times to check it out.

One circle around the field showed that Nancy hadn't arrived yet, so I entered the landing pattern. Fortunately the wind was coming from the west and the Crash Crew was stationed on the east side of the field. It would be much easier to sneak Nancy aboard. I saw the fire truck, the wheel watcher, and the Corpsman.

After two touch and go landings, a car the color of Nancy's drove around to the west side of the field and stopped. The next time around, I landed and rolled out. I slowed down and turned off onto a gravel taxiway and came to a stop at the far west side of the airfield.

I looked over toward the Crash Crew. They were almost invisible from this location. I waved my handkerchief a few times and Nancy shot through the bushes. With the plane idling, I jumped out and helped her into her seat and buckled her up. She gave a thumbs up. But she didn't know yet what an uncomfortable ride this would be with her back against that hard seat back and no parachute for a cushion.

I explained that we'd be taxiing toward the Crash Crew to take off so she should bend her head as low as possible so they wouldn't see her. I hopped in and taxied toward the east end of the runway toward my take off point. Before getting too close to the Crash Crew I wheeled around, gunned the engine and within seconds was airborne.

We headed west and flew over Mobile Bay. Nancy was having a good time. A half hour later, time was running out and we just barely had enough time for Nancy to get out. We entered the Brewton landing pattern and behold, white flags with large black "Xs" had been laid on the approach end of all of the runways. A huge red flag waved on the field flagstaff to signify that the field is closed. But the Crash Crew was still there. Very suspicious. I didn't say anything to Nancy yet.

I went into a near panic and headed east toward Saufley. Nancy was still in the plane. I was due to land within twenty minutes and it would take at least fifteen if I made a beeline now. Inside those five minutes I had to figure how to get Nancy out. My eyes scanned the ground below as my mind worked to find a solution.

Mercifully, I saw a huge pasture below. Without hesitation, I set up to land there. I touched down rather roughly on the grass and shut down.

"What are we doing here," she yelled.

"Nancy, you gotta get out."

"Get out, hell! I don't even know where I am. Which way is my car?"

"That way," I told her and pointed toward the west.

"You're crazy. A million snakes out here."

"You can't go back with me. I'd be in all kinds of trouble."

"You got me in this. You're getting me to my car. I'm not getting out of this airplane."

"OK, but you and I are finished."

Nancy didn't know it, but I headed back to Saufley, got in the traffic pattern and called for landing. The tower responded in its normal way for me to continue my approach to the duty runway. Things were looking better. Now I had to figure some way to get "sweet thing" off the plane secretly.

My touchdown was beautiful. I wished my instructor had seen it. I continued to the taxiway and turned off toward the parking area. Still things looked good. I pulled into the chocks, thinking that since it was the last flight of the day, I'd tell Miss America to stay in the plane. I'd leave the canopy unlocked, and tell her to sneak out later when the traffic died down.

What I didn't know was that an MP jeep had trailed us at twenty yards since we turned off the duty runway. If I had seen it, I would have realized that, in addition to the driver of the jeep, an officer and three MPs were aboard. Gee whiz, they only need one.

Nancy was a basket case. I was dropped that day. I got a letter of reprimand along with my dismissal slip. Since there were no damages, I guess the Navy felt dismissal was enough. As for Nancy, the Navy drove her to her car.

First Solo

The first big step when a student reaches Primary Stage is getting that first solo behind him. Every one who has ever soloed remembers that day. It's the perfect example of a perfect day. After flying that solo, the student will sometimes meet his friends at the club, maybe have his shirttail cut off and, whether summer or winter, be thrown in the pool. The new hero is expected to treat those friends to a round or two and afterwards they cut off his tie.

This story comes from, Chris, a former flight instructor who should have known better. He knew that a relationship between a flight instructor and a flight student should be all business in the cockpit and non-existent out of the cockpit. And it is especially difficult to control when friendships existing prior to entering flight school enter the mix when one is an instructor and the other a student.

> *I should never have let this happen. We graduated from college at the same time. We married our college sweethearts upon graduation. He was from Iowa and I was from Florida. We entered the Corps immediately after graduation. Since then, neither had seen the other. In 1953, we met again in Quantico in the same USMC Basic School class. We immediately resumed our friendship. He and I and our wives, the four of us, became inseparable.*
>
> *I requested flight training and left for Pensacola. He selected Infantry and was ordered to Korea with the First Marine Division. While he was overseas for that year, his wife visited my wife and me in Pensacola two times. I think she felt a little closer to him when she was with us.*
>
> *Upon graduation from flight school, the Marine Corps ordered me to remain at Pensacola as a plow-back. I went to Instructor School and afterwards I was assigned to Squadron 18 at Saufley Field, the so-called all-Marine squadron.*
>
> *Immediately upon my friend's return from Korea, he applied for flight training and was accepted. What a happy day when they arrived. The four of us began to renew our friendship. He began preflight at Corry Field. I continued teaching in Squadron 18.*
>
> *After Preflight, he went to Saufley Field for Primary and was assigned to Squadron 18. I knew enough not to request him as a student. One Thursday I saw on the schedule board that my friend had finished his A-11 flight, so I asked the scheduling officer to let me fly his A-12x, that infamous OK-for-solo check flight, the very next day. He agreed.*
>
> *As he did his air work, I saw he was a very good pilot. We flew to the outlying field at Foley and shot some touch and go landings. It was then time for me*

to stop the plane, shut down our pretty little yellow T34B, climb out of the cockpit, and launch my friend on the traditional two go-arounds, all by his lonely self, to finalize his OK-to-solo.

Foolishly, the day before, we had arranged for our wives to drive over to the practice field at Foley. Bad idea. We would be doing our groundwork there. More bad idea. I also had arranged for the field crew at Foley to provide the wives access to the field so my friend's wife could take pictures of his first solo flight for posterity. Awful bad idea.

I saw that the girls were in place so I gave my student a few words of encouragement and joked to him about smiling for the camera. He taxied down to the far end of the taxiway and made his turn onto the active duty runway. He shoved the throttle forward and began his roll.

Suddenly I saw the wind speed increase dramatically and change its direction by a full ninety degrees. I prayed that my friend would pick up on that wind and adjust for an excessive cross wind.

He didn't. As the plane rolled faster, the T-34's wings began to create more lift, lightening the weight on the wheels. The plane began to crab and bounce at an angle off to the left. In no time the wheels were bouncing and skipping across the deck, sending the plane off its intended track by forty-five degrees.

There was nothing I could do but watch. I saw my friend's wife, standing ten feet off the left of runway about three hundred yards away. She apparently saw him veering to the left of the runway, so she wheeled and took a picture as he sailed barely ten feet overhead, still skidding and struggling for altitude.

How he recovered, I'll never know and neither will he. Fortunately, he milked the stick to gain a little lift, then lowered his nose a bit to pick up speed, over and over again, continuously gaining bits of precious altitude, all as intuitively as if he had flown for years. I constantly wiped sweat from my face, praying and hoping, wishing I could undo what was underway.

All those prayers seem to work when his plane stabilized and began a slow climb. He exited the flight pattern, most likely to try to compose himself. In a few minutes he re-entered the pattern, and followed the rules for the student's first solo on an A-12X flight. He flew two touch and goes, then approached the third time for a full-stop landing.

He shut the plane down. I saw him hunch over, I'm sure in relief and exhaustion. I stood on the wing, reached in and gave him a couple of pats on the back. He and I were covered with a river of sweat that accumulated through that episode.

Back in the squadron area as I completed the write-up for his check flight, I gave him as high a grade on that flight as I could. I said nothing about the peril he faced. But had there been a provision for instructors grading themselves, I would have given myself an unsatisfactory. No one ever had such a lesson as I did that day.

That night the four of us went to Martine's Steak House. Before splurging on a "prime rib for four," we each said a prayer, an earnest prayer.

Attack on the Liberty Ships

Our jitter boxes activated as the big day approached. An A12x (OK for solo check flight) had to be flown to a tee. Instructors were tough. They didn't want to unleash an unsafe student.

Our buddies knew when and who'd be flying A12x flights. How? There it was — in plain view — on the big black status board. And heaven forbid if we were to fail that check ride. A big red "hold" message would be clipped to our names and visible to the world.

And more to the point, we had to keep in mind that every minute of the next flight, the A13s solo, we'd be on our own. This time the back seat would be empty. Nobody would be back there to say, "Here, I've got the plane. I'll show you once more."

Darryl submits this story, but claims that Curt should share half the blame for getting them into the following shenanigan:

Curt and I became good friends during Preflight. We were Naval Flight officers, both having gotten our commissions through NROTC. When we moved on to Primary Stage at Saufley, we requested assignment to the same squadron. The Navy complied.

It also happened that our A12x check came up on the same Friday afternoon. We both passed and learned that each would fly our A13s solo at 0800 on Monday morning. We had no idea that fate was setting us up for one wild Monday.

During dinner that night, we celebrated with an extra glass of wine and planned the weekend. With solos coming up Monday we could manage less weekend study, so we elected to further rejoice by a trip to Destin Beach the next morning.

The beach was almost empty when we arrived. The look of the sun promised a scorching day so we set up for an early stint on the beautiful white sands then we'd retreat to the Hog's Breath, our favorite place, and some brew. Besides, nice little cuties usually breezed through a-trolling, and we might let ourselves be snagged.

At noon we each ordered a beer and a Hog Monster Burger. We took a seat at a large table, where two guys were seated. We discovered they were Air Force pilots from Eglin Field, so we struck up a conversation.

Our new friends congratulated us on our A-stage success and seemed pleased that we would be soloing Monday. Our conversations went one way, then another, and finally one of them suggested we should have a little fun Monday because it might be a long time before we'd get that chance again. They mentioned that the old ***World War II Liberty Boats*** *were docked at a remote spot on Mobile Bay. They also mentioned that Air Force pilots often ran low altitude gunnery passes on them in F-4s as they returned from training missions.*

"Is that legal?"

"Kind of. Just stay away from the shipping and dock areas."

The captain explained how we could set our radio to a different radio frequency and the Navy would never hear us. That sounded good. They also suggested we use some seldom-used frequency.

"I'm thinking of one. It's one-two-one- point-five," the lieutenant said. "Few people use it. You'll have it all to yourself."

I jotted down "one-two-one-point-five" and said thanks.

"By the way, what time is your flight?" asked the captain.

"Zero eight hundred. Monday."

"Have fun guys. We'll be thinking about you. Good luck."

They left. Now if we could get this plan together.

Monday morning, we took off and headed west. We cleared Saufley Field at 0830 and changed to 121.5, our "secret" frequency and radio checked. It worked perfectly.

Now we had to find the boats. I took the lead. Having had no experience in flying formation, Curt stayed well back. Dead ahead I saw Fairhope and descended to 700 feet and turned north. There I saw the boats.

I keyed the mike. "Leader to flight. Target dead ahead. Follow me." I nosed over and at 300 feet I voiced a rat-a-tat-tat sound over the radio then pulled up. In a few seconds I heard another rat-a-tat-tat and I knew my wingman had joined the fight.

Still transmitting on our "private" frequency, we enjoyed several minutes of amusement, and then decided to head to Foley Field, shoot a couple landings, and return home. "We don't want to be late, partner," said Curt. "Go to channel four now."

Back at Saufley, we landed, checked in our planes, and walked to the ready room. My instructor came over and asked how my flight had gone and what I had done.

"Fine, sir. Some stalls, a couple of spins, and a lot of landings at Summerville and Foley."

The squadron commander came over, and with a glare, handed me an official looking note. "Your so-called partner has one too," he said.

I looked at the note. I was ordered to report immediately to Operations to see a certain lieutenant commander. "What's this about, sir?"

"It's not addressed to me, Darryl. Best you guys go find out for yourselves."

I got topside to Operations and a yeoman led me into the lieutenant commander's office. Curt was already there standing at attention at the officer's desk. The commander asked us how we enjoyed our first solo flight.

"Fine, sir." Curt mirrored my answer. I began to think maybe this might be a positive meeting.

"I'd like to play a recording for you two heroes," said the commander. It started — a complete replay of our attack on the Liberty Ships. Over ten minutes worth. My mind sank. This had to be my last day in the Navy.

The officer began an atrocious attack. During that tirade we learned that the frequency we used, one-two-one-point-five (121.5), was an official "guard" frequency, recognized worldwide and used only in case of ***actual emergencies****. Our conversations were probably heard by every aircraft, every ship, every radio within hundreds of miles, whether it was Navy, Army, Marines, Coast Guard, police. "Even the frigging Air Force," he said, emphasizing Air Force.*

AIR FORCE! It hit me. Those damn Eglin pilots led us down the garden path. A trap. Bet they listened in. I hope to live long enough to run into them at the Hog's Breath.

Curt and I escaped with only a "headwork down". Somebody was looking after us. Gratefully our previous records were clean. After a re-fly and a re-check, we continued in the program. And stayed clean.

CHAPTER 10

Basic – Stage I

We had finished Preflight, then Primary, and about three-fourths of us were still in the game. We were learning the Navy way. Our flight instructors always cautioned us that even though we had accumulated well over enough hours at this time to apply for a civilian private pilot license, we didn't know crap. We lacked training in navigation, instruments, night flying, cross-country, and most of all, were not yet blessed yet with a significant dose of common flying sense.

Primary Stage had been our first air test. It was thorough. It taught us we could aim a plane down a runway, keep it fairly close to the centerline, determine when we reached V1, then V2, lift off into the air, fly from point A to B to C, and sometimes find our way home without a mayday call to the tower. Simple!

Of course, it was important that we would successfully place that plane on the ground without breaking it. Some didn't learn. The Navy watched and dropped a few. Some decided they didn't want to remain in the cauldron. No harm there. Plenty of other openings in the surface Navy or in other walks of life.

You Ain't Crap

Shockley, a Marine student officer, was accepted for flight training out of Marine Basic School in Quantico in 1962. He submitted this story:

We had finally made it through Preflight, then Primary, and finally were starting Basic Stage. We were proud. It was only ***Basic Stage I,*** *sometimes known as Baby Basic, but we were there. We hadn't yet flown a single Basic Stage flight but our attitudes had changed (and that was bad). Basic was definitely a cut above and all of a sudden some of us began to consider ourselves somewhat as members of a super race. Simply moving from Saufley's forty some hours of flight time to Basic Stage ground school at Whiting must have done that.*

One the first day of ground school, we sat in a classroom waiting for our instructor, a Marine Corps captain. We would soon learn what we were. We snapped to attention as he stomped into the classroom waving his swagger stick. He slammed it hard on the rostrum. I thought it surely had broken.

"Listen to me!" he ordered. He began to tell us what he thought of our misguided self-evaluations we had submitted and how off-base we were. For over ten minutes he continued to put the fear of God in us.

At the conclusion he said, "You people ain't crap. Not yet anyway. You might think you're an almighty bunch but it's time you got off your high horse and get with the program. You have a long way to go before you can even think about membership in the finest flying organization in the world."

We came back to earth. We began to understand. We got the picture of who we were. It would be a long time before we'd rise above the level of crap.

In Basic Stage I, we plunged into learning the Navy way, the professional way. Basic came on fast. We worked 12 to 14 hours a day; sometimes more. Sadly, more of us would make an exit. The first sub-phase "Transitioned" us from our Primary Stage philosophies that had taught us to be safe and moderately competent. Now, we moved into a newer, more powerful aircraft. Basic Stage was moving us forward and onward.

We became familiar with our new plane and moved to "Precision" sub-phase where we learned to hold altitude to within 50 feet instead of double that. We learned to spot-land and the spots were small. Suddenly we progressed into a more seriously demanding brand of "Acrobatic" flying, with lower deviations in altitude, azimuth, airspeed and mistakes.

Next in line was "Basic Instruments," a sub-phase where we flew every flight under a hood for an hour or two, covering 400 miles and not seeing a square foot of our track. Most students remember these flights as tedious, monotonous and even boring. Our instructors kept riding us. We were chided for being off speed, off altitude, and off course. We didn't perspire, we sweated. And we began to realize why instructors back in Precision were so persistent.

Interwoven within the Basic Instrument flights, we were thrown "Navigation" events: day and night navigation, dual and solo navigation. This sub-phase afforded many thought-provoking hours. We were actually accomplishing something. Maybe it was that just traveling from point A to point B to point C excited us. We weren't under some dreary white sheet hearing the dull drone of the engine and waiting for the instructor to key his mike and say, "OK, take a break."

We learned about IFR navigating in an even more interesting phase, "Radio Instruments." The Navy still believed in each of us, so we'd climb aboard, get under the hood, and realize we were proceeding to and landing at real locations, not just tooling around over the Gulf of Mexico at 25,000 feet. We recalled Basic Instruments and were thankful to those instructors for pushing us and chipping at our mistakes.

Next we entered an incredibly interesting sub-phase – "Formation." It seemed as if we were finally Flying Navy. The jovial mood around the Formation squadrons pointed up a high state of morale. We flew seven or eight two-plane hops. We learned why Precision instructors demanded that we stay on altitude, course, airspeed, and not wander off willy-nilly. That was even better demonstrated when we moved from two-plane to four-plane. Wow! Being completely enveloped by several aircraft within two to three feet of us, made us hope the others had learned a lot also.

Our training to this point began jelling in our minds when the fewer and fewer of us who still remained in the program, entered "Gunnery." We changed our training location back to Saufley Field, amongst the T34s. It was exciting to taxi down the flight line of tiny T34s so the Primary Stage students could look up and see us in our T-28s. Talk about haughty.

The Navy actually had the fortitude to hang four automatic weapons loaded with live .50 caliber tracer ammo under our wings, send us out over the Gulf to climb to 18,000 feet, to enter a dive, and try to riddle an aircraft-drawn target pulled by our instructor at 7000 feet, without shooting that poor sucker down. We'd make five runs on each flight and improved on each run.

Instructors were brave. They would constantly harp on students to keep their guns spouting tracers at forty-five degrees or more from his flight axis. No fear, we were learning. We were getting control. Gunnery was what Naval Aviation was all about. Very soon, those fortunate students still in the program would transfer to Advanced Training. We were ready. But first a few stories from former Basic students and instructors.

Pitching Pennies

In flight school, weather was the chief factor that determined who could fly, when, and where. Devices the Navy called "condition lights," looking much like lights up and down small town streets, hung high inside each training hangar, indicating the up to the minute launch status. Aerology, the base weather office, would constantly study the weather and, if needed, would launch a weather plane to verify weather conditions and report. The lights would be set to green, yellow, or red according to current conditions.

Green signified that the weather was clear. All flights, solo or dual, could launch as scheduled. This condition had a nickname; launch everything.

Yellow indicated that duals only, no solos, were permitted to launch. Sometimes in a marginal yellow condition, advanced Basic students were authorized to fly certain types of solo flights. The nickname for yellow was duals only.

A *red* condition light meant no training flights were to depart. This red light signaled that some nasty weather is close by and was known by its code name, hold everything.

In hold everything conditions, weather planes or flights specifically approved by operations, would depart. During these times all students were expected to remain in their assigned squadron area ready rooms, study, and be ready for a condition change.

Ready rooms were furnished with student lockers, large conference tables with chairs on which students could study awaiting their flights, and small tables where instructors and students could brief or debrief flights privately. Portable six-foot high partitions marked the area boundaries for each squadron.

Most training squadron commanders mandated that during a hold everything condition, student pilots were to study and be ready to fly as soon as cleared. Some commanding officers enforced the rule of study more strictly than others.

Timothy tells his story of how not to defy the laws of the squadron commander:

> *In 1956, I was a Naval Aviation Cadet. I loved flying. I had finally made it through Primary Stage at Saufley. My first two flights in the incredible T-28 had been great. I began thinking I'd switch to flying as my life, even though I actually wanted to follow a career in journalism.*

Sometimes flight students hoped for a hold everything weather condition to rest and regain composure after days of the stress of fly, fly, fly and study, study, study. They'd get up a Bridge game or poker. I liked Bridge since I was a teenager but it was not on our study menu. Some of these Bridge addicts worked out a scheme so that playing Bridge could look as if they were studying.

Our squadron commander, and I use the word loosely, was a young and self-important Naval officer who had just been promoted to lieutenant. We secretly called him Heinrich because if you paired his arrogance with his moustache, he was almost a double for the one and only Heinrich Himmler, one of Hitler's trusted and inhumane generals. Our Heinrich went to the Naval Academy at Annapolis and after graduation was sent to Pensacola to fly Navy. After he completed flight school, he spent a tour at NAS Jacksonville aboard a carrier.

After one sea tour, he was sent back to Pensacola for duty as a flight instructor, where he was assigned to Basic Stage at Whiting to take over a new and understaffed training squadron, Squadron 9. That squadron, at the time Heinrich assumed command, had very few students. Since Heinrich was the senior officer in Squadron 9, he became the squadron commander. I had the misfortune of being assigned to Squadron 9.

One rainy morning all the training squadrons at Whiting were working under condition red. Only the weather planes were flying. The skies looked as if they would stay that way at least until after lunch.

At noon, morning flyers would depart for ground school and afternoon flyers would report to the flight line.

I should have been studying. I had just gotten to Whiting and had a long way to go. But why not have some fun?

A Bridge game was underway. I thought about getting up another foursome but I thought it might draw Heinrich's attention. I went to the head and thought about a game we used to play in college. In our dorm's bathroom some of us guys would play pitch-a-penny. Players would stand ten feet or so from a wall. One by one we'd flip a penny toward the wall. The player whose penny landed closest to the wall would win all the pennies for that round. No one ever won or lost more than a few cents. Remembering, I began to practice in the head.

A couple of my friends came in and saw what I was doing. They joined me and before long, two games were underway. Too bad. Sounds of laughter increased without our realizing it. In walked Heinrich. "What the hell are you guys doing?"

"Playing pennies, sir," I said.

"Every person in here right now is on report. This is dereliction of duty. You're supposed to be studying. Sign this paper. You will stand before the Board. Get to your squadron areas on the double."

Heinrich did put us on report. The flight board, headed by a commander, brought all of us, including Heinrich, in at the same time. They took Heinrich's statement and dismissed him. One by one the commander took our statements. He told us that our squadron commander was technically correct in his position. "But," he said, "Sometimes we have to give a little and soft pedal rules. You guys have been working hard. Here is what we'll do. I am giving each of you a written assignment. Within a week from today, deliver your completed work to me in a sealed envelope. I will review them. Afterwards, I will talk to your squadron commander and explain that your have completed your tasks and no further action is necessary."

From that day forward I stuck to the rules. Which one of those officers would I die for?

Bad Hair Day

We were there to fly but the Navy never allowed us to forget that neatness counts. Except under battle conditions, Marines live and die by spit and polish. Navy and Coast Guardsmen certainly understand neatness, cleanliness, shiny shoes and close haircuts, although maybe not to the extent as their highly disciplined cousins.

Many students in the flight program will remember Admiral's Review, an event to be attended by everyone. Student flight officers, Navy, Marine, and Coast Guard from previous duty in the field or sea duty were old hands at keeping themselves sharp. They came from military academies, NROTC, ROTC, OCS, Marine Corps PLC or OCS, or maybe from the ranks. For at least a year they had worn the uniform of their country and were accustomed to spit and polish. As such, the Navy did not require them to participate in most of the weekly, mostly Saturday morning, inspection required of Cadets – guys just out of college – guys with lots to learn about the ways of the military. But the Admiral's Review was an all-hands event.

On the last Saturday of every April and every October, promptly at 0800, we would stand at attention in formation and the admiral or his staff would walk by every student, inspect them, and call out infractions of neatness and proper uniform to accompanying aides. Actually, squadron commanders would have already conducted very detail and picky inspections prior to the admiral's so this was only a procedural task.

At the end of the cursory walk-around inspection, the admiral and his staff would take their places on the reviewing stand. The NAS Navy Band would sound off with patriotic Navy, Marine, and Coast Guard strains and the admiral would command, "Pass in review." Read below this memory that Robert, a former Marine student flight officer, shares:

> *A very tough Marine major had recently reported to NAS Whiting Field. Some students said his face was the mold God used to create bulldogs. He had served as an enlisted infantry platoon sergeant in Korea. For a series of acts of heroism and leadership under fire, he was awarded the Navy Cross, the Purple Heart, and a battlefield commission with the rank of first lieutenant.*

He returned to the States for special ceremonies and recruiting events.

While home, he applied for Naval Aviation flight training. Now, with all the hype awarded this Marine, who on God's green earth would turn him down? He was accepted and fought his way through the year and a half, graduated, and returned to Korea to fly aerial gunfire spotter planes. The war ended, and by 1958 he had progressed to the rank of captain, was promoted to the rank of major, and transferred to Pensacola for duty. There, he was assigned as Gunnery Squadron Commander at NAS Whiting Field.

As a Marine first lieutenant and former Infantry officer, I entered flight training and in 1960 was a Gunnery flight student when this tough Marine Major took over the squadron.

We worked hard. On the last Friday of April, I flew three hops. The third was a night run, so I didn't get in until after 2030. The next day was Admiral's Review. I needed a hair cut, but barber shops closed at 1800. None of my friends in the squadron would even think of cutting anyone's hair.

I got home at 2130. My wife, Marjorie, re-heated supper and we talked of my dilemma. I could barely imagine enduring the wrath of my Marine spit-and-polish squadron commander if I were to fall out for Admiral's Review parade and inspection without a recent haircut.

Marjorie thought hard. "I'll do it," she said. "I'll use the clippers I use on Tweedle." Tweedle was her tiny fluffy poodle and my mortal enemy.

"I don't want stupid dog's hair mixing with mine." I said. Then I realized that Marjorie does a good job on Tweedle so I relented. In good old Marine tradition, I said, "OK. Any port in a storm."

I got out the bed sheet that Marjorie always unfolded and spread when she cut Tweedle's coat. Tweedle yelped and ran as she saw the familiar setup and disappeared to one of her unknown hiding places somewhere in the house. I moved the sheet within view of the dining room mirror and set up a chair. Marjorie plugged in the clippers and the motor's whine sent Tweedle racing to still another of her secret places; this one probably deeper and even more remote.

Marjorie began to cut. I closed my eyes. She suggested that her best approach was to run a few cuts down the center of my scalp and work outwardly on both sides, again and again, working down to my neck. I stole a look in the mirror. Not so bad, I thought. It'll work. I felt better.

Marjorie turned off the clippers so we could get a good look. "Won't be long now," she said, laughing at her joke.

Not bad, it'll work! I thought again. So far, she had cut a nice four-inch swath, a quarter-inch in depth, down the length of my scalp. It was evenly done. It'll work.

Marjorie took a sip of water and switched on the clippers to continue cutting. As she positioned the clippers, we heard a nasty grinding noise. The clippers came to a dead halt. I don't know what I said, did, or thought, but I was sure of one thing — this was not good news. And it was after 2300 now.

"Your clippers quit," she said.

"My clippers? Really?" I said in disgust. "Not my clippers. I've never used them."

Marjorie lined up in front of me and sighted down my head. "You've got a reverse Mohawk," she said, trying to hide laughter. "Some of the kids I knew at Auburn used to have them."

My mind went into a tailspin. What can I do? In frustration I blurted out, "Gotta take emergency leave!"

Marjorie shouted, "How about Nancy and Sig?"

Sig and Nancy. Two of our best friends. He was a Navy LTJG flight student. He and I transferred into flight school the year before and ended up in the same Preflight class. And Nancy always cut Sig's hair immaculately. Isn't it amazing how, in a frustrating instant, wives usually pop up with perfect solutions?

I called Sig and explained. Minutes later, after he finally had the decency to stop laughing, said, "Sure. Come over." As I hung up the phone, I could hear him still cackling.

Marjorie stayed home. I didn't blame her. I got to Sig's at midnight. Nancy, probably in trying to hide her giggling, fiddled with the clippers. Sig drew a beer for me to complement his.

The next morning after the inspection and parade, that usually ultra-tough, feisty Marine major, popped me on the shoulder, and bearing a devilish grin, said, "Good haircut, Bob."

Sig was always known to "let the cat out of the bag" if it wouldn't produce faulty consequences. Later, when I checked out of the squadron to move on to the next flight phase, I asked the major if he had known. He bore an even more devilish grin with, "Of course not."

Nancy and Sig were true friends. Marjorie and I caught a lot of ribbing from that night until we graduated — true friends. None better.

Gunnery

Yes, Gunnery! The word alone summons excitement in the heart of every Naval Aviation student pilot. They could imagine cruising at 4,000 feet in their Sopwith Camel over the Marne River in 1917. Looking down, they'd spot none other than Baron Manfred Von Richthofen, the great German ace, 2000 feet below, returning to base in his blood-red Fokker DR-1 tri-wing.

They'd begin to imagine: "He's mine," they'd shout as they'd roll into steep dives from six o'clock high. Down, down they'd dive toward the Red Baron. With white scarves streaming in the slipstream, they'd yank back the cocking arms of their dual .303 Vickers machine guns. Closer. Closer. Now from 100 feet aft, they'd lock and hold the triggers hurling hundreds of killing rounds into the target.

In seconds, smoke would surge from the cylinders of the Baron's Thulin rotary engine. Down, down in flames the Baron would snake. With scarves waving, they'd yell to the heavens, "Bravo! The death of Major Hawker has been avenged." What imaginations!

Gunnery was the most exciting phase of the Flight Training's Basic Stage flight training syllabus at Whiting Field in the 50s and 60s. G-1, the first flight in Gunnery phase was the only dual flight, an indoctrination flight. The instructors loved it. It was probably because those teachers got to let their hair down and act like kids again. Flat-hatting (flying too low) was forbidden at any time in Naval Flight Training, as well as everywhere else, but since gunnery flights were conducted at an extended distance offshore and at higher altitudes, few observers were around to see or tell or care about these Pappy Boyington wanna-bees.

Something new had been added to those wonderful T-28s or the T-34Cs since students flew them in previous training phases. The Navy crammed Norden gunsights into the forward cockpits. It almost seemed as if a mummified dog lay stretched in front of the pilot, occupying more than its share of space. Moreover, a forward see-through panel of the canopy now contained a strange-looking 6" x 8" see-through prism. This removed nearly half of the pilot's original front panel view.

The theory of the Norden Gunsight, simply stated, was ingenious. In flight, when the pilot switched on the gunsight, a gyro began to spin at a high rate of speed. A *reflector* emitted a high intensity spot of light in the center of the prism.

If the aircraft banked to starboard in a level turn, the gyro caused the reflector to move the spot of light to the right, an amount relative to the rate of turn of the aircraft. And likewise, in a port bank, the spot moved to the left.

If the aircraft entered a dive, the gyro caused the reflector to move the spot of light down the prism, an amount relative to the rate of descent of the aircraft. And likewise, in a climb, the spot of light would move up the prism.

If the aircraft entered a diving turn to the right, the gyro would cause the reflector to move the spot of light down the prism and to the right, an amount relative to the combination of the rate of turn and descent of the aircraft. Ingenious.

Therefore, the theory of the Norden gunsight is that the spot of light on the prism of the gunsight screen should always be displayed so that if the pilot kept the spot on his intended target as he fires, he should get a kill. This did, however, assume that the aircraft was flying in balanced flight and the pilot practiced enough.

The usual Gunnery flight in Basic consisted of an instructor pilot in his T-28 pulling a 15-foot long canvas target sleeve, and four gunnery students, each in their own aircraft with two .50 caliber machine gun pods hung under the wings.

Each plane's ammunition was coded with a different color (red, green, yellow or blue) by dipping the business end of the .50 caliber rounds into waxed paint. When a round was fired and passed through the target sleeve, some of the colored wax bled onto the sleeve around the bullet hole. At the end of the flight, the instructor would drop the sleeve on the recovery field. As soon as the students returned, they would break their necks to race to the sleeve and count their hits. Such competitive spirit existed especially when Operations began to post each day's results on a status board.

To begin the flight, the instructor met his fledglings in the ready room for a few minutes of briefing. They discussed specific points about this flight. After a few questions and answers, the instructor and the students would head for the Parachute Loft. After some nutty amenities by the guys in the Loft, such as, "If it don't open, bring it back," the instructor and his chicks would head for the flight line to sign out there assigned planes.

With aircraft preflight complete, the instructor would take off and head for the target pickup area, a secluded runway just north of the field, to snatch the 15-foot long white target sleeve from the deck. He would scream in at 320 knots and 20 feet altitude. With his carrier hook down, the instructor would lock onto the 100-foot long cable tethering the target, yank his control stick full back and fly straight up, full power, to 200 feet then push the nose over. Checking back, he would see that he had acquired his target sleeve then pitch to level flight. Once he gained full control, he would head toward the Gulf, southward over a vastly remote area. Never would the Navy want a target sleeve to fall and damage property or persons.

By this time the instructor's four Gunnery students were in the air, having flown west from Whiting, over the smokestacks of the factory at Cantonment, Florida. Here they turned south to cross the beach at the water tower at Pensacola Beach. On each flight a different student flight leader would escort the flight to the Gunnery area, about 30 or so miles due south over the Gulf.

The instructor, in his tow plane, kept in contact with the student leader and directed the flight to perch altitude, 15 to 18,000 feet, to a hold position over the clear blue waters of the Gulf. Now the fun would begin.

Most students reaching this phase of training were looked upon as students most likely to advance to Advanced Training and soon graduate and receive their Navy Wings of Gold.

However, it could be said that even though these students were good pilots, the instructor in the tow plane might be seen as putting himself in a bit of jeopardy. From his viewpoint, the students above him were good, although relatively new pilots, in planes armed with dangerous and deadly .50 caliber bullets. And the poor instructor guy sat right out front.

Students at times have a tendency to fixate on the target banner to the point that they forget the remainder of their surroundings; even the tow plane. At times when planes in the area switched to the Gunnery flight's frequency, they might hear a few expletives and maybe even a few nasty or spiteful directions. The Naval services do an outstanding job of selecting students for Naval Flight Training. Their job is to examine students and determine which will most likely make the grade and especially ones who will not shoot down Gunnery instructors.

Gunnery in Basic enters the syllabus about a year after the student checks in for Preflight. Most of the drops have occurred. The remaining students have now acquired about 170 hours of good hard Navy training; training on a considerable higher level than anyone ever gets in a lifetime of tooling around in a Cessna

172 for thousands of hours. A flight instructor said to me during an interview for this book, "These guys are damn good, but don't tell them I said that."

For example, consider my preflight class of 31 students. We checked into Flight School and spent over a year and a half going through Preflight, Primary, Basic, and Advanced. In the Preface to this book, I mentioned the Navy Commander who spoke to my preflight class. He told us that statistically half of us would not graduate because of various reasons. Thirteen of my original class of 31 did not graduate. And not one of them shot down a Gunnery instructor.

Some flight students begin to see the handwriting on the wall at various stages during the year and a half. Others just plug along, sometimes oblivious to their fate of being dropped. Flight School was designed so that every stage was more difficult than the stage before. Many of these faltering students would request a proficiency evaluation even before an instructor broached the topic. Some were in the "He who knows not and knows not that he knows not" category. Another student might feel that no matter how poor I am, I can get better.

Four Out of Five Ain't Bad

The next story comes from a retired Naval Aviator, a Marine and once a Gunnery instructor in Basic Stage. He said that once the average student reaches this stage, he has settled down into a state of seriousness, but not Hamilton, one of his charges. Read this sent in by John, Hamilton's instructor of years ago:

> *I was a flight instructor and had recently completed the Instructor Indoctrination Course. This gave me a license to teach student Naval Aviators at Whiting. Hooray! I checked in with my training squadron and was assigned to fly a few live Gunnery hops in the back seat with some instructors as they led their flights of students on the road to becoming fighter pilots. I thought I would really enjoy flying high in the sky, out over the Gulf, and in charge of experienced students. My turn in the barrel*

was approaching. But if I had known what was in store for me with Hamilton maybe I would have run for the hills.

I walked into the ready room of Squadron 8 on a cold, November morning. The best part of it was that the skies were as clear as I had ever seen them. Visibility would be no problem.

I looked at the board and saw that my flight, 8-C, consisted of four students, and was scheduled for 0930. I checked my watch. It was now 0815 hours. Where could they be? I'd like to have a chance to meet them, get to know them, where they came from, and let them meet me. Sure, it was only 0815 now, but students are supposed to be in their squadron well in advance of their scheduled flight. Of course, 0815 is well before 0930, but...

I heard laughter coming from the recreation area, so I walked over and saw four students at two Ping Pong tables. I stepped over to them and spoke. They politely came to attention and greeted me. I asked if they were 8-C.

The student whose nametag indicated Hamilton, spoke up very politely, "Yes, sir. Sorry sir. We didn't know if our instructor would be here today."

"I am your instructor. Don't you think it would be a good idea to stand by in the ready room just in the remotest chance that I might want to talk to you?"

"Yes, sir." It was Hamilton again. He seemed to have become the spokesman for the group. The five of us walked to the ready room.

This was to be their G-4 flight, meaning that it was the preliminary to a grand and glorious live ammo flight, G-5. I had looked forward to seeing four hot shot students on one of their last flights before heading to advanced flight training.

We had a set-to, an old fashioned set-to, where I really laid down the law about how the remainder of this phase would be run, even though they had almost finished Basic. Then we briefed this day's upcoming flight. As we split up, I gave them rendezvous instructions. They departed with instructions to sign out their planes and be at that rendezvous on time and on altitude; not early or not late. I signed out my aircraft, took off, picked up the target and headed for the rendezvous point over the Gulf.

They performed with perfection. I noticed that Hamilton seemed to be flying with an apparent lower level of regard for safety than I would have desired. I was towing the banner at 12,000 feet. They perched at 18,000. When Hamilton would come off the perch, he'd dive at too small an angle, almost lining up with me. I couldn't even see him. Thank God this was not a live fire.

When we landed, I pointed out to the flight, and especially to Hamilton, that tomorrow was live shooting day and to set a goal to approach the banner at not less than forty-five degrees.

The next day, they waited in the ready room before I arrived. That's good. In my briefing for the flight, I congratulated them for their presence and stressed again that the angle of approaching the banner should be on the money at not less than forty-five degrees. "Hamilton," I said, "Try for forty-five degrees. I thank you; my wife thanks you; my kids thank you. I do not want to get wet, especially in the Gulf of Mexico on a cold morning like this, or neither do I want to die."

The first run was flawless. I saw those .50 caliber tracers zipping by with ample clearance. It looked as though most of those guys must have had hits. Then Hamilton came in a little slim. I warned him. The next run was better.

"OK, guys," I said over the mike. "One more run. We'll go home after this pass so completely empty your magazines. I want to see some good tallies when they post our scores." I told them to hesitate on the perch for a second or two and think about what they were going to do before they started their dives. "Visualize your run."

Hamilton was last on the perch. The first three students flew beautiful patterns and I knew a lot of hits went through that banner. Then came Hamilton. As he called, "two oh five, off the perch," I knew I'd better keep my eyes on him. Suddenly I lost him in my peripheral. And I couldn't find him in my rear view.

Then I saw tracers zipping by in a steady stream barely off my starboard, continuously and too close for comfort. It was Hamilton. But where was he? The tracers kept coming. All of a sudden another flood of tracers shot by, followed by a huge shadow not thirty feet above me and less than a twenty degree approach angle.

Hamilton zipped in front of me and off to the right and did this victory roll that only the Red Baron could perform. He pulled his nose to the vertical until he fell off in a hammerhead stall.

"OK, eight-C, let's go home," is all I could say. I began to think about what to do. No way did I want to down Hamilton for this. He flew a beautiful pattern. Most of the time I was not in jeopardy. As we walked to the ready room, I talked to him. I scolded him but explained that since the others didn't see his maneuver, I wouldn't report it. "Ham," I said. "I know one day you'll either be a superb Naval Aviator or die young."

Scores? Navy records show that not all G-5 students hit the banner. A four or five is good. Ten is excellent. Ham had 47. Damn! Pray tell how, God?

Strip Dance for the Crash Crew

This story comes from Arney, a retired Naval Aviator. At the time this story happened, the late fifties, Arney was a LTJG and, as he said, had just been sentenced to serve as a flight instructor at Saufley Field. On this particular day, Arney was functioning as a Landing Safety Officer with the ground crew at Brewton Outlying Field, north of Saufley Field. This meant he was supposed to watch as students' aircraft approached for landing. In the event of an apparent unsafe condition, such as wheels up, Arney or someone in his crew would wave them off. The most common signaling method was the firing of a red flare. But students were always cautioned to abort a landing any time they saw any signal that could possibly be considered a warning message from the crash crew Arney's story:

> *I had just returned from a carrier-based attack squadron in Korea and you can't believe how happy I thought I was to be assigned to plush duty as a flight instructor at Saufley Field. This was one of my first days, so I assumed that because I was a new man on the totem pole, I had the lowly task of serving as Landing Safety Officer and Safety Officer at Brewton Field, working with the Crash Crew. I didn't get a single flight hour for the whole week. It was our duty to eliminate headwork accidents.*
>
> *It had been a boring day. For some reason, very few students were selecting Brewton Field on that day. After more than an hour of boredom, I decided to stretch my legs with a walk around the outside of the safety truck.*
>
> *About thirty feet away from the truck, I spotted a large reddish mound looking very much like a pitcher's mound on my high school baseball field in California. Unaware of the danger, I stepped on the mound and took a few steps. Little black things. These things I found out later were fire ants! They had warned me about them. Ants by the thousands swarmed and began marching straight at me.*
>
> *Before I could protect myself I was covered head to toe and being stung unmercifully. The guys in the truck didn't hear my yelling and had no idea of my plight. I knew I had to get out of my clothes. I tore off my flight suit. That only showed me that hordes of these fiery creatures had already made it to my skin and were intent on devouring the rest of my body. Next I yanked off my boots and saw these horrible creatures had attacked my ankles and feet. I was in total distress.*

I tried everything. I flailed my arms like a windmill against my body and brushed off every ant I could see. I stomped the large clusters of the ant colony. Then I noticed the crash crew saw me and were laughing for all they were worth and without a bit of help.

Then I saw two planes approaching. They were in the landing pattern very close together. The first one had just turned final, the other was off his tail, although too close, turning final. They probably saw my ridiculous frolicking and, thinking I was frantically waving them off from a perilous landing maneuver, jammed full throttle and pointed their noses skyward. No two planes had ever flown so close to each other without occupying the same airspace.

Next they made a low-altitude flyby and couldn't interpret the situation. They must have been so confused that they thought they'd better get out of there so they left the area.

Later I learned that these ants were not just usual everyday ants. These insidious creatures were called fire ants, a vicious insect common to the Southern states in the Gulf area. No other name would suit them better. But a trip to sickbay for some high-powered medicine and two days of recuperation put me back on duty. I learned never to tangle with those creatures again.

Olenole's Solo Work

The idea of flying Navy was intriguing to James. Getting in was not as tough as he thought it would be. That accomplished, he was sure it would be a walk in the park to graduate. Preflight was a breeze for him. He progressed well in Primary Flight. Ground school was tough, but it was coming along, too. He found he even had some time for liberty on the weekends. It was 1951. James was 20 and he was ready to scatter some MiGs from North Korea.

During Basic Stage he began to discover that flying wasn't a second-nature thing for most students. He soon understood that very few good things are easily attained.

Here is Jim's story:

> *At last! After those sixteen weeks of Preflight, I was ready, along with most of my fellow classmates, to break the bonds of earth! I had no doubt that my aerial abilities and agilities would amaze my flight instructors and send my fellow students cowering into the shadows.*
>
> *I was bewildered when I learned that we'd have a full week of ground school before our first flight. The brass seemed to taunt us by guiding us past the flight line of the SNJ (Air Force T-6 and the Army Air Corps BT-13). We endured it by staring out of the windows during classes as we watched those J-Birds landing and taking off and imagining how much better a job we could do. And we believed it.*
>
> *Soon there was the big day. I remember it. It was Monday morning as we walked into the flight hangar. I looked for my squadron flag, the flag of Squadron Six. I found it quickly. Some of my classmates were also in Six, but the remaining members were distributed among the other five squadrons.*
>
> *I looked around the area. There was the Big Board that ground school instructors talked about. It looked like a grammar school blackboard, but a tad more complicated. I soon learned to digest the information on that board: when I'd fly, who was my instructor, what I'd be doing.*
>
> *Our flight syllabus in Primary Stage consisted of 19 flights (A-1 through A-19). A-19 was the solo check. I knew I'd be there in nothing flat.*
>
> *Things began well. I thought maybe I actually was cut out to be a Naval Aviator because I could find the two grass one-mile-square practice fields. The Navy had it figured out well. If we were practicing full-stop emergencies, we'd use the left 1760 feet width of the*

field. If we were shooting touch-and-goes, we'd use the center 1760 feet width of the field. All full-stop landings were done on the right 1760 feet width of the field. These sections add up to 5280 feet. One mile. Smart huh?

A high-visibility tetrahedron indicated the active runway direction: north, northeast, east, southeast, south, southwest, west or northwest directions. Space was ample.

A couple of other fields in the area had other specific purposes. Foley was used for only cross wind landings. Pace was always used for pre-solo checks.

All in all, it was not a bad plan. It seemed geared for safety. But shortly I found flying more difficult than I had expected. But why not? Those who wear the Wings of Gold are a cut above the rest and must buckle down. We all knew that.

As time went by and training progressed, I did realize that I had a kind of knack for making good landings. I felt that was good because at least I had a better chance of getting on the ground safely.

Some of the maneuvers at high altitude were not so easy for me. I remember learning that a plane stalls for one reason. It has lost its required lift. Spins are like they say; a spin is a stall gone bad. The word spin didn't make me feel at all secure. But I soon realized if all the others could do it I would do it.

We'd be confronted with other things like loops, barrel rolls, Immelmanns, crazy eights, tail chases. These were just part of the game. Whatever leads to those wings would make me happy.

About half way through the A flight syllabus, we were scheduled to receive a progress check by flying with an instructor other than our own daily tutor. Of course, there were other times when our hero might not be available, and we would fly with a different instructor. It was different getting used to someone else. He was our mother hen.

My instructor seemed pleased with my progress, especially my landings. He pushed me hard so that I would feel relaxed when it came time to be checked. He was certain that I could pass the test.

Soon the day came for that mid-A check. I was assigned to a short, stocky instructor who didn't talk; he growled. He was not a commissioned officer. This was a surprise. He was an enlisted pilot. He had been a Navy chief who received his designation in the latter part of WWII. I secretly named him ACE.

After introductions and hangar instructions, ACE sent me out to get my parachute, inspect the assigned aircraft and be ready to fly. I preflighted and got in the cockpit. I strapped in and started the engine. There were a few checks the student could make before the instructor arrived.

By the time ACE arrived, we were ready to go. To say that I was not worried or overly concerned would be stretching the truth quite a bit. I just hoped that the guy didn't show. Or maybe I would get sick and start throwing up.

I knew ACE would be using a magnifying glass. The little clipboard on his kneepad was used for comments. No doubt he had a full list of maneuvers he would check.

The plane captain pulled the chocks, and I began to taxi out to the take off area. Since the SNJ had a tail wheel, I couldn't see over the nose. This meant that taxiing consisted of multiple "S" turns. I could only see as the plane rotated from dead ahead. This was necessary to clear the area ahead of us.

We arrived at the take off end of the runway and waited our turn. The takeoff was fine although I had some thoughts about whether I had pleased my checker.

We proceeded to the assigned areas for our "high work." I suppose that the acrobatics were acceptable along with the stalls and spins. I even made a good high altitude emergency practice landing. It was all coming along nicely.

Soon, we were off to Foley Field for cross wind landings. These could be a tough challenge at times, but I appeared to pass the test.

These check flights were longer than a regular flight since we had to do all of the procedures. Normally we would be gone 75 minutes for a regular flight. The check could last up to two hours. "Now take me to Pace field", he said. So we were off to the supreme test, but I didn't have much of a problem with landings. I entered the pattern cleanly. I came around on my down wind leg in preparation for my first touch-and-go landing.

The SNJ is a tail dragger and our procedures said to flare out about one or two feet above the ground and then stall the aircraft. At this distance the plane would simply drop on all three wheels and roll straight ahead. Once in control, the power would be added to proceed down the runway and become airborne once again.

ACE had me making my approach and flare quite a bit higher than the height to which I was accustomed. I became unnerved by the procedure change. He specified that I flare closer to six or eight feet.

My landings were not as smooth and controlled as the ones that I had been doing. I became rattled. I didn't feel right about it.

Then, on one more try to make a more acceptable landing I came around with the added burden of having my heart in my throat. ACE was "chipping" away. Too vocally. He couldn't let me do it my way and then offer constructive criticism. He never let up.

"Try one more," ACE growled. I started my approach. This would tell the tale. Again he growled at me to level off in my flare. It was so high that the prescribed stall became a precursor to a "crash."

And that's what we did. The SNJ stalled so high that it couldn't be controlled between there and the ground. The plane fell off to the right and we went rambling down the runway in a completely uncontrolled ride.

ACE grabbed the controls and tried to settle down the bucking bronco. We finally made it to the end of the runway before getting the thing straight and slowed to a safe stop.

Boy, was he mad! I won't repeat what he said. The one good thing was that, as an instructor, he was officially in control of the aircraft so the accident would go against his record. He growled about that.

Our inspection found that the wing tip was bent and there was grass in the bonding wire of the flap. The plane was not really in bad shape but ACE was.

Had I been quick on my feet I might have said, "I guess that means that I don't get to solo, right?" But I didn't. I was just too scared.

After an official inspection he flew the plane back to Whiting with hardly a word. I followed him to the board where he gave me a down.

He gave me the routine debriefing but seemed awfully glad to be rid of me. I saw it as if my approach and landing was not satisfactory he should have taken the controls and prevented the fiasco. Oh, well.

My record had been quite good up to now, so the administrators decided to give me a few "extra times" and schedule another check flight. Surely I could straighten out that little problem with a few additional flights.

My extra times were mixed with my own instructor and others. It went well and I was soon put up for a re-check. This would be the big day.

The next guy seemed a little more reasonable than ACE. He didn't have a chip on his shoulder. He wasn't a head banger. What a refreshing change.

My work at altitude was fine. The cross winds were adequate. The high altitude emergency was on the mark, and I plunked it right in. Now for the practice touch-and-goes and to a full stop. I would then let the instructor out and make a few take offs and landings of my own.

Okay, here came the practice landings, but all of a sudden I developed a swerve on roll out between landing and take off. The landings were fine, but how did that swerve get in there? I had been doing it on every landing. This instructor was trying to be my guardian angel. He was giving me more and more landings. We were already over time on the flight.

I just couldn't do it, but he didn't want to let me get kicked out of the program. But, what could he do. He wouldn't want my life on his hands.

He decided to pull away from Pace and head home. I was crushed. I'm sure there were many tears in my eyes. My heart was pounding. I could see myself labeled as a failure. It wasn't a pleasant feeling.

Once comfortably airborne and heading back to Whiting, he asked if I thought I could straighten that swerve out with one more flight. I told him that I was sure that I could. He must have felt that also.

When we returned to the hangar, this prince of a fellow picked up the yellow sheet on which the maintenance discrepancies were notated. He "downed" the plane for a failed intercom just so I wouldn't get a down. Imagine! We were out over two hours, and he wrote up the yellow sheet saying that the plane had a technical problem that prevented us from completing the flight.

What a guy! I could have kissed him right there.

I did go out the next day or soon thereafter and get another check. This time I performed like a veteran.

My landings were great with no swerve. After three or four touch-and-goes I made a full stop landing on the runway and pulled around to position myself for a take off and also let the instructor out.

I was on cloud nine. Here was the big thrill I had long awaited. The instructor was out and monitoring me via radio.

All I had to do now was line up and take off. I did, and the feeling once I became airborne was indescribable. I was free! I'm like a bird! This was what it was all about.

But I realized I had done only part of it. It seemed easy taking off. Now I had to get this thing down, and by myself! What if I pick up the swerve again? What if I stall it too high on my landing flare?

My exhilaration suddenly took a back seat. It was again time to get to business. I just had to make this landing and the others safe and smooth.

No problem. I plunked that dude right in there. I did two or three more the same way. Then the instructor called me to make a full stop landing and pick him up.

Once back in the plane, he took over and flew us back to Whiting. He knew that I had put all I had into that flight and those landings. I needed a rest.

Hey, really... it was nothing to it!!!

CHAPTER 11

Basic – Stage II

Flight Simulators

The Link Trainer, the first of the Navy's flight simulators, came into being in 1929 when Edwin Link found that he could not afford the high costs of learning to fly. He wanted to continue his learning so this engineer-type young man spent 18 months inventing and building a flight simulator.

His first version actually looked like a wooden airplane, having wings and a fuselage. He mounted the device on a universal joint so it could rotate 360 degrees and was powered by small electric motors connected to pneumatic drivers. The trainer itself was mounted on a table about three feet high.

His Pilot Trainer worked so well that the Army Air Corps saw that it might be a solution to their rash of 12 airmail pilots killed in a little over two months because of their lack of knowledge of how to fly in poor visibility. Link changed the design by removing the wings – they were of no commercial value. He improved the controls so that as the pilot moved his controls the Link performed more like a real aircraft.

The Army Air Corps signed a contract for six pilot trainers at a total cost of $21,000. His company expanded rapidly as the war years loomed closer. They began to produce a machine every 45 minutes — over 10,000 units during the war. These trainers saved hundreds of thousands in fuel and flight time costs and permitted pilots to receive much more training than normal.

The pilot trainers manufactured by the Link Company served well even though they were somewhat crude. Pilots and students who trained in them into the 80s will remember the flops, the droops, and the slumps as the units made turns around the map courses.

Here is a story by Layton, a training assistant:

Battle of the Link Trainer

I was a Naval Aviation Training Assistant at NAS Pensacola for the years of 1955 through 1959. I spent many boring hours sitting at a table outside Link Trainers while students and pilots followed our map courses. I'd watch a little ink pen make tracks over a paper map. This mark would show the precise route of the student's flight.

As students arrived for their flights, I helped them into the trainer, briefed them on their clearances and their communications. Once their flight began, I constantly changed their routes. All the while, the trainer would turn left, right, up and down as the student followed instructions.

The universal joint used in the Link Trainer was not precise. Sometimes the machine jerked around and the pilot swayed in the cockpit. Often a pilot yelled for me to quit yanking him around. I'd yell back but this was all in fun.

One day a student came in for his first ride. He had a tough time getting the feel but finally was doing a good job. I pressed him by giving him harder routes to take.

The universal joint on his trainer box was especially loose, causing his ride to be bumpy and slinging him back and forth. It was humorous because he kept poking verbal jabs at me and I'd jab back.

Suddenly his trainer made a hard ninety-degree turn and I heard the universal joint of the trainer snap loudly. Before I realized it, the entire Link trainer had fallen from its table to the concrete deck, making a huge noise.

We pulled the student out and I knew he was hurt. A corpsman rushed up, looked him over and called an ambulance. Several days later we heard he was returning after a stint in the hospital and that he had had a concussion. We were told that all indications were for a total recovery.

The next day he was on our schedule for a warmup flight. When he walked in we had a surprise for him. The staff and students gathered around as we pulled back a cover from a table covered with snacks. Our Lieutenant presented him with a mock Purple Heart Medal after which we sang, "For He's a Jolly Good Fellow."

My White Lights are Red

Cheating may be described in many ways. There is light cheating, medium cheating, and downright wholesale cheating, such as cheating your way through college. Eddie's story tells of a time he did some light cheating, but even so, he didn't get away with it:

When I began Basic Instruments at Whiting Field in 1976, I was a typical Student Naval Aviator trying to balance flying the aircraft using correct procedures, developing an efficient instrument scan, and all the while trying to stay collected under the instrument hood.

I was not entirely successful and I knew this phase might be an obstacle to me as I tried to kick the ball through the graduation day goal. I began to look for ways to make things easier so I could bide some time and maybe get the hang of this thing, or even art, they call "instrument flying".

Find a crutch, I thought. I would soon be flying my next BI flight at night and I had to find a crutch. So the night before, I painstakingly copied all of the procedures for the flight onto my kneeboard. I wrote in red ink so I could spot it easily, or so I thought.

We took off just at nighttime and climbed to work altitude. My instructor told me to get set up because we would be running through the whole program, start to finish. Afterwards we'd come back to redo where I might be having problems.

"You have the aircraft," he said. "Follow the program."

I looked at my clipboard. Damn! I couldn't see a single crib note. I knew I had written it down. Then it hit me. I had not taken the time to learn the cockpit lighting scheme in the T-28. If I had, I would have known that the lighting in the esteemed T-28 was red. The red light in the cockpit preserved my night vision but rendered my kneeboard cliff notes invisible and useless. This scenario made for some very anxious moments that night under the hood. I confessed to my instructor. He saved my butt by talking me through the hop. That demonstrated that in the future I should make a definite point of finding out what I was facing.

I finished my successful quest for my wings and spent a couple of tours aboard a carrier and another on shore duty. Ironically, for my next duty assignment I was transferred to Whiting Field as a flight instructor. Also as luck would have it, I was assigned to the very same basic instruments squadron. But now we were flying the newer T34-C. They had been upgraded to the new low-white instrument lighting! The new low-white gave better vision but does not affect night vision.

Stearman Surprise

Acrobatics or aerobatics — however you call it, it has always been a major part of Naval Air Training. Acrobatics teaches a new flyer to handle an aircraft positively, gently, and precisely. The checklist for acrobatics is exact. Every item on the list must be completed. If not, unexpected results might happen. This story comes from Fred, a Naval Aviator who left college after two years to try to earn his Navy Wings of Gold:

> *I started college in 1938 to study to become an accountant. Newspapers began to fill with talk of Hitler, a madman in Germany, invading small countries in Europe. The same message came out of Japan about Tojo. Both of them seemed to want to own the world. When I finished two years of study, I applied to the Navy to be an aviator. I couldn't believe I passed the flight exam and even was more surprised that my flying seemed to be good enough maybe to get me all the way through. I look back now and remember one flight and what a stupid stunt I pulled.*
>
> *I liked the Stearman. Just being up there with the wind blowing over my face was exhilarating. I made it to acrobatic stage. That was what flying was all about. My instructor was a gentleman. He worked with me very hard if I had a problem. Sooner or later I'd get the picture and both of us were happy.*
>
> *But when I'd do something stupid, he'd be all over me. I did something stupid one day. Really stupid. He told me to go out to the plane and get ready for our next acrobatic flight. I preflighted the plane as well as any one had ever done. I waited at least ten minutes, so I got worried and asked the lineman if I should go look for him. He said no because he'd be here.*
>
> *He did. He came rushing out and told me to start the aircraft and taxi out to the takeoff area. I did a good job. I learned to zig zag when taxiing so I could see on both sides. This was my third acrobatic flight and a little nervous about whether I knew some of the procedures well enough.*
>
> *At the take off area I stopped the plane and turned into the wind and began the run up. Everything looked good so I gave a thumbs up to my instructor. That was the only way I had to communicate. We were using the*

Gosport. I had the listening port and he had the speaking port. I couldn't talk to him.

"OK, Fred," he said through the Gosport. "When your acrobatic checklist is complete, roll it. Get out to the area and climb to four thousand."

I started my takeoff. It was a great run. I lifted off with no errors and began a climbing turn toward the Alabama line. I got over Foley Field and my altimeter showed four thousand feet.

"All ready?" came through the Gosport. My thumb went up.

"Then give me a half-Cuban eight."

I nosed over to pick up airspeed then pulled up sharply, holding about three Gs. At the top I was inverted so I began to release backpressure on the stick. The G-forces began to drop off.

That's when it happened. The Gs hit zero, then went negative. I expected to feel the familiar sensation of hanging in the safety straps. It didn't happen. Instead of hanging, I tumbled out of my seat. I watched my upper wing go by and my plane disappear high above me. It was ghostly.

I got the picture. I had fallen out of my plane. What do I do now? I felt air rushing past me. I reached for my ripcord and pulled it. I expected it to open and thank God it did. In a few seconds I heard the chute streaming out of its sack. In another few seconds I felt a horrible yank on my harness. My chute was full and I was slowly descending toward the earth. The instructors used to tell us to look up. If your chute is full, thank God you're OK but you might not know it yet. I said a prayer.

My instructor had flown back to check on me and was circling a few hundred feet above. He was yelling something but I couldn't make it out. I just gave him a thumbs up.

I landed hard and hurt my ankle. I hopped around on one foot and collected my chute and made an "X" on the ground with it. I remembered that's what to do if you can't walk out.

They picked me up in a jeep. My ankle was badly sprained and I went to the hospital for over a week. When I came back to duty I got ripped over the coals by my instructor and my squadron commander. They told me they had tried to keep me from being dropped from flight school. They couldn't because it was a rule.

I had the choice of being discharged and going back to school or to stay in the Navy with my choice of duty. I stayed and was sent to school to learn to drive Higgins boats, hauling Marines to battle the Japanese.

My only consolidation is that I have heard that there have been more than a few cases of this brand of stupidity. I'm just thankful mine ended with only a minor injury.

Sir, Is This an Up or a Down

In my many conversations with my great friend, Marine Colonel Dodenhoff, we talked at length about the methods some instructors use to enlighten students as to how they fared on a just-completed flight. I would never have realized that this technique he mentioned was rooted nearly 3000 years ago in old Rome. Here's another submission by Dode:

Not long after I began flying Naval aircraft (1942), my instructors would tell me that I got either a thumbs up or a thumbs down." I learned quickly that an "up" was a satisfactory flight and a "down" wasn't. An up sent me to the next flight in my conquest. A down was another matter. I'd have to re-fly the flight successfully and sometimes twice. Once was to retrain and the second for the record. A down just messed up a student's life. I'd think this was the end of me. When I'd have to re-fly, an added amount of nervousness would always creep in.

Some of us began to ask around where the Navy ever came up with the names ups and downs. We thought maybe we should ask one of the best instructors in our class if he could answer the question. Commander Allen was not only a great flight instructor but he was a serious history buff.

It was a rainy day in November and everyone was sitting in a ready room waiting for the weather to clear. We asked Commander Allen about the names up and down.

"I understand it probably started in old Rome, close to 3000 years ago, and the Vestal Virgins." We all stared at each other. "Vestal Virgins," he continued, "were an important part of Roman culture and Roman government. The government appointed six of these. Just as the name implies, these women were virgins. They had to stay that way for their thirty-year term of office. If they broke and engaged in sexual relations, they were buried alive, because they had broken the law. The Virgins, being high in government, could not be punished with the spilling of blood. Being buried alive accomplished that.

"What this has to do with ups and downs? Remember the stories about the sword battles in the Coliseum? Two fighters would fight it out until one collapsed. The winning fighter would lay his sword against the neck of his downed opponent and wave his hand toward the Vestal Virgins. That was the signal for them to give a thumbs up or a thumbs down. A thumbs up meant that the fighter had put up such a good fight that his life should be spared."

Somebody piped up, "Commander, if the Virgins gave a thumbs down, guess that sword would go right down guy's throat."

"You said it guys," said the commander. "Let that be a lesson. We don't like giving downs. Our home life is so much better when you guys fly well and we can point that thumb straight up."

Don't miss the best of Dode's great remembrances as he tells about, "The Avengers of Pearl Harbor," in Chapter 7 of this book.

Suck Less Next Pass

When Bunky was in flight school at Saufley in the middle 1990s, one of his ground school instructors thrived on repeating this old flying adage, "You can teach a monkey to fly but he can't make the voice reports." Crazy thing, but this author remembers the same saying in flight school over thirty years earlier. That theory could be suitable in many cases and scenarios and stated in other ways.

The Navy way to teach flying is to pre-prepare, pre-prepare, pre-prepare, then when all seems right, train a little more. With proper training of either brand new or longtime seasoned pilots, they will be ready to cope with adversities thrown in their direction.

Bunky's most memorable instructor was a Landing Signal Officer. This LSO's job was to instruct T2C Buckeye students as they flew Field Carrier Landing Practice on a concrete runway for a couple weeks before going to the ship for the first time. He'd stand at the approach end of the facsimile flight deck, paddles in hands, to signal flight faults. Needless to say, his call sign was "Paddles" during training flights.

Bunky tells this short but solidly appropriate story which students who have endured that training will understand:

> *My Field Carrier Landing Practice LSO at Corpus Christi, Howie, had a strange method of ringing our bells as we flew FCLP practice. A jokester he was anyway, he used his comedic talent well. It was certainly effective with me. Occasionally one of us would be too high or low in an approach; too far to the left or right; or maybe too fast or slow.*
>
> *On one occasion, I was in the hot seat. The first time around, Howie yelled at me, something like,"301, you're coming in hot. Shave power."*
>
> *"Wilco."*

The next time it happened, he made a sharper admonition. "Get with it, 301. You overshot every wire. You're in the water. Swim hard and watch for sharks."

By the third time, he took his more sterner but laughable Howie approach to the problem. He yelled in the mike, "301, Paddles, over."

I came back, "Roger, Paddles. 301. Go ahead."

"Dammit, 301. Lousy sucking pass, 301. Why don't you just plain suck less next pass."

"Roger, Paddles. 301. Understand plain suck less next pass. Wilco."

And by golly, I did suck less next pass, not that I fully understood what he was trying to say. But I guess I figured it out. I got an up on the flight.

Carrier Fright

In 1970, Leroy Banks was assigned as a Carrier Qualification instructor in VT5 at Saufley Field flying that ever-superb T-28C. Some CQ squadrons were still based at Whiting while the transition progressed to Saufley.

Saufley had been home to only Primary students for quite a long while. These Primary students were new to the game of flying. For many, this was their first experience at any kind of flight. As could be expected, the CQ students were old hands, having over a hundred and fifty or so Navy flight hours inked into their logbook. What bragging rights! Some CQ students disparagingly referred to the newbies as babies. It wasn't unusual to see a flight of CQ students with superior attitudes taxiing their T-28Cs past the T-34 line, all the while cycling their tail hooks up and down and glancing at the newbies.

The T-28C, with that five feet long, distinctive tail hook, set these guys apart; far apart above the babies flying the T-34s. CQ was a high point during a student's flight training days. It was the last phase before moving to Advanced training.

Nevertheless, even these hot shots sometimes required a bit of instructor assistance. The first phase of CQ was Field Carrier Landing Practice (FCLP). An outline of a portion of a flight deck was marked on the runway. Students flew a pattern around the FCLP field, landing and taking off, again and again and again. The object was to get the feel of a *low and slow* pattern, then touching down in a precisely small target area, more like a spot, an ink spot almost. At first, the altitude and airspeed for FCLPs were about 300 feet and 160 knots. As students progressed toward the conclusion of FCLPs, altitude and airspeed were reduced to a precise 178 feet high and 128 knots, and lots of flaps and power.

Instructors rode with FCLP students on their first few flights. This was a new world. Students had, as the old saying goes, "a long way to go and a short time to get there." After that, instructors watched from the sidelines, looking for unsafe or erratic performances. They showed no hesitancy to pull a student and assign him an extra hop or so. If the student couldn't readily improve, he was ejected from the program. After several FCLP flights it was time to head out to sea to test their skills on the ship.

A month before the events of this story occurred, a CQ student had approached the carrier on one of his final landings aboard ship, when something went awry. He was coming in too low. The LSO (Landing Signal Officer) standing beside the meatball, saw the condition of the approach, gave him a wave-off, fired a red flare, and transmitted the abort to the student. The student pilot did not respond. The wheels of the T-28C hit squarely against the deck top, tearing the landing gear from the plane, and sending the plane cartwheeling down the deck in a horrific flaming fireball. Needless to say the student died in the accident. Word of this disaster "spread like wildfire." Here is Leroy's story:

> *I foresaw good things for one of my students in my CQ flight. Frank had served as a navigator with the rank of LTJG aboard an APA troop carrier ship in the Sixth Fleet, a very responsible assignment. Nobody called him by name. Everyone referred to him as Satch. Why? I didn't know but maybe he did.*
>
> *Having made three Mediterranean cruises, Satch accumulated an extremely good military record in the Fleet. So far during flight school, his flying history*

foretold great things. His FCLP work was excellent. He would hit wire number three more than half the time. I could see he was grooved on the meatball. He was excited to fly carrier qualification flights aboard the boat and especially to progress to Advanced Stage, as he said, hopefully to jets.

Early that Monday morning, the bus picked us up at Whiting and delivered us at 0700, just as the sun began to appear over the horizon. As we preflighted the planes, he talked incessantly about his future in Naval Air.

Before we boarded, I went over a few major points that he should remember. The last point I stressed was that he could do it. I told him that since he only needed six satisfactory and successful solo arrested landings to qualify, he'd probably get them by Wednesday.

We took off in our flight of four aircraft and headed south over the Gulf to intercept the ship. I handed off the plane to him because this would help to relieve built up stress until our rendezvous with the boat. That was great plus with him. Such confidence, I thought.

Ahead, we saw our target boat and such a beautiful sight it was. Our four-plane division joined up and one by one we got the call from Primary Flight Control to make our first visual pass. "Do not land. First pass is a look see," they ordered. We had descended to 300 feet and I told Satch to fly the first pass and I'd talk him through every turn.

"No," he said. "You fly and I'll follow what you're doing." I thought that was strange, but I flew the pass.

Next, Flight Control ordered us into the landing pattern at 300 feet high and 140 knots. We would be spaced well apart. Soon it was our turn. I told Satch to watch; that I'd talk myself through and he could watch and listen.

I didn't hear a word as we flew the pattern. I told him to get the picture of our attitude and the meatball. I snagged number three wire with perfection. "That was a good one, Satch," I said. Deck hands cut us loose and we taxied to take off.

I asked Satch if he'd like to do the honors. He declined, so I took off and climbed to 600 feet and waited for Flight Control's order. When it came I told Satch he had the plane.

"I can't, sir."

"Can't?"

"No, sir. I feel sick."

"How? What's wrong?"

"Just sick. I'm about to throw up."

I dropped out of the flight went back to the field. We talked. I asked Satch if he needed more FCLPs.

"Yes, sir."

Because of his superior record, the squadron flight administrator permitted Satch to drop back two weeks. I worked with him. Again, his FCLP work was perfection. I felt that he could recover, so we sandwiched in with the next group going out to the ship.

Same thing again. I even sent him to the shrink. The doctor called me and asked if some recent incident might be the problem. I thought back and remembered that less than a month before our flight went to the ship, a student came in too low, crashed, and died instantly.

With that bit of background the doctor felt that Satch could not take control of the aircraft at the ship when we were at a low, even a safe low altitude, where he might see the water staring him in the face.

The Flight Board dropped him. In my last conference with Satch I told him I'd help him any way I could in his exit review to help him stay in the Navy.

I don't know if Satch stayed in the Navy. I can say that his removal from flight school was a terrible waste. An even worse waste would have been if he had to leave the Navy.

Wild Table Turn

A formation may be defined as a configuration, an arrangement, a structure, or maybe a pattern. Clouds in the sky form patterns. Everyone understands this. In this book, we are talking about the arrangement of a common group of aircraft taxiing on the ground or flying in the sky.

A non-commissioned officer knows that when troops are to be moved from one location to another, no technique is better than to collect the unit into formation and give the commands, "Fall in. Right face. Forward march." Now we have our platoon or squad moving in the desired direction and fully under control.

By issuing other commands, the unit is directed toward the destination. Then, the NCO says, "Platoon, halt. Fall out." The platoon has arrived speedily at the intended objective with a minimum of commands.

Even when I was in the first grade the bell would ring, signaling the end of recess. Our teacher would have us line up in two columns and remain in place. She would say, "No talking. Follow me." We knew we must follow Miss Dennis' orders or suffer some consequence.

During World War II, one mission of the US Air Corps was to engage in long range strategic bombing over Japan and Germany. Sometimes a thousand or more bombers and fighter escorts would take off and head for places like Ploesti, Berlin or Tokyo.

Can you imagine sending a thousand aircraft toward Tokyo, individually, in a disorganized swarm with no one in command? Of course not. Then how did commanders maintain control of such an armada? Certainly some element of command and control.

Maybe formation? According to the number of planes, the entire flight might be broken into air groups, further into squadrons, flights, divisions and sections. Every commander would receive his orders from the next level of command, then report to his superior, who would report up the chain to his superior, and so forth. Every aircraft commander, every pilot, knew just how he fitted into this huge cluster of aircraft called a formation.

And so, at an early stage in his training, after becoming somewhat competent in the basics of flying, the Naval Aviation Student is introduced to one of the important elements in the military way of flying; formation. No longer is he buzzing around the countryside looking for make belief tanks to bomb, or cruising low along the beaches watching the pretty babes sunning themselves.

Now he learns to fly the Navy way, in orderly clusters of two or more planes, in formation. But formation flying is not just a few aircraft tooling around in the vicinity of each other.

Rules are established to smooth the process of flying. One pilot is in control. He makes decisions, or may delegate responsibilities. He files, or may delegate, the filing of the flight plan. He may delegate tasks but he may not delegate the ultimate responsibility for the flight.

During the formation flight, the planes usually remain in close formation in controlled airspace, but may spread into a loose formation when enroute to a destination. But every pilot in the flight is aware of the location of each plane. In flight school, formation flying will become the normal for the Naval Aviation student and later the Naval Aviator.

Invariably, people play stunts on each other. This story comes from a former Navy pilot, via YouTube. It happened when he was a Formation instructor at Saufley Field. The SNJ was the plane of the day. His story will be paraphrased from a combination of a taped account currently running on YouTube and a conversation with one of the students in the story.

> *A new instructor, "Vern," joined our Formation Training Squadron after completing the Flight Instructor Indoctrination Course. Any time a new instructor reports, other instructors size him up for "fitness." Word had it that he seemed a bit arrogant. This could have been because he was a "plow-back." In other words, he might have just completed flight training himself and for his very first duty station, was assigned to NAS Pensacola, as a flight instructor. (Plow-backs sometimes seemed*

to assume a superior and arrogant attitude. Just ask student who had one for an instructor.)

A couple days later, Vern finished his local squadron indoctrination and was assigned to fly a four-plane formation check flight with four students he had never met. He assembled the four students for an extensive preflight briefing. He finished by selecting one student as leader and gave him the altitude and the coordinates of the rendezvous point. "I'll see you there," he said.

The four students and the flight instructor checked out their chutes from the parachute loft and headed toward the flight line to find their assigned aircraft.

The plot unfolds here. The four students quickly and secretly exchange places with four seasoned Formation instructors and in minutes have their engines running and are airborne.

Vern reaches the rendezvous point. He looks around and finds no chicks. He makes several radio calls to locate his students and finally locates something that resembles a gathering, a covey, even a crowd, that appear to be strung out for a half mile laterally and a thousand feet vertically. He closes and sees the plane numbers belong to him.

First, Vern attempts to restore order. The "students" do not maintain position. The lead changes constantly, and by accident. In echelon right formation, the leader turns right — directly into the path of the formation. And vice versa.

On and on, the fiasco continues with improper hand signals and unannounced altitude changes. After an hour of accomplishing virtually nothing, Vern coaxes the wild formation back to Saufley for landing. They pass over Saufley on a westerly heading, and make the break. This would have put them in ideal position for landing on runway 9, the duty runway. But without warning they immediately execute a beautiful second break, putting them in perfect position for landing on 27, exactly 180 degrees off course.

At that point, Vern tries to reorganize his flight and in doing so, becomes so exasperated that he makes a request of Saufley Tower that on the next pass, they should shoot down all four planes and rid the world of "a scourge too appalling to bear."

This goes on. The flight lands. As they roll out and switch to ground control, Vern ordered his chicks to meet him in the ready room immediately for a meeting. The meeting was far different than he expected. When he walked in, he heard the recorded replay of the flight and sees the senior pilots still carrying their chutes. The students who had relinquished their cockpits join this meeting and the entire ready room is caved in a state of laughter.

CHAPTER 12

Advanced Training

Ejection Seat Trainer

The Navy prepares pilots for predicaments that no one hopes will occur in flight. To do this, they train in devices that simulate dilemmas and how to recover, all the while being safely and solidly on the deck.

Over the years, ejection seat trainers have changed dramatically as much as planes have. In the Fifties, the device looked a bit like a child's playground slide with a misplaced ladder.

The slide portion was actually a long track, angled maybe fifteen degrees from the vertical. Instead of a curve at the bottom, like that of a child's slide, it was dug into the deck for a support.

Two one or two-inch diameter steel pipes, serving as a track, extended about 20 feet into the air. The theory was that the trainee would climb into the ejection seat and an attendant would press the firing button to set off a controlled explosion under the seat, propelling the trainee up the track.

Explosions were created by inserting an explosive charge, looking much like a small spent artillery round, into a firing chamber at the base. A firing officer would estimate the weight of the trainee, drop in the calculated number of explosive bags and press the button. This powerful shot would heave the ejection seat containing the strapped-in student a distance that depended upon the firing officer's calculation of his weight. Typically, the student would rocket two thirds the way up the track, after which the safety officer would pull a release lever to lower the seat and the student to the ground.

Ejection Seat Fiasco

Ray tells this story about his ejection seat training at Advanced Training in 1951:

> *It was the near the end of June, and a hot day in Texas. My class lined up for our first crack at the ejection seat trainer. The safety officer didn't want any problems on his watch and worked slowly and meticulously. The students got restless.*
>
> *Our class clown, Gene, began milling around the mechanism picking up some strange yellow bags from the ground and putting them in his pocket.*
>
> *I walked over to him. "What's that, Gene?"*
>
> *"Explosive stuff, I think."*
>
> *"What are you going to do with it?"*
>
> *"I don't know. Probably have a big Fourth of July bang-off down at the beach."*
>
> *Where Gene was our class clown, we also had Redmond, our class bitcher. Any time things didn't go his way, he'd bitch, bitch, bitch. There's no other way to describe him other than a major bitcher. We tried to shut him up. We never knew when it would roll back on us.*
>
> *Gene was still nonchalantly collecting the yellow bags for his huge celebratory show on the upcoming Fourth of July. Now it was Redmond's turn in the bucket and he made his way toward the ejection seat, bitching*

about how hot it was, how thirsty he was, how he was going to fly S2Fs anyway so why was he doing this junk.

As Redmond was being strapped in, I kept my eye on Gene. He casually inched his way toward the firing mechanism. About the third time Redmond said "dammit," Gene secretly dropped a couple of yellow bags into the casing.

The safety officer set the weight gauge, inserted his calculation for the number of bags to put in the firing chamber, loaded it, and gave the all clear. *The firing officer pressed the firing button. BBBOOOMMM! What a shot! Sounded like the whole world blew up. It brought workers streaming out of the maintenance building to see what happened. Redmond was still strapped in the ejection seat, but was stuck on the track with his head two feet above the very top of the track. The ejection seat had cocked to about twenty degrees to one side. Another erg of power would have sent him over the top. We all looked at what had happened. Only Gene and I knew the whole story.*

They brought in a front-end loader and retrieved the bitching Redmond. Only this time he had something to bitch about. Mechanics checked the device, did a few trial ejections and gave it an OK. Engineers concluded that Redmond was much lighter than any of the other students and the firing officer had miscalculated the number of explosive bags to use. He was cautioned to use better judgment, especially when a light student was on the track.

Nudist Colony

This story would not have been written until recently, and certainly not until a former instructor and his student had retired from the Navy. Otherwise both might have been called back to stand before a Navy disciplinary board. That was the opinion of the former instructor who submitted this story. In any case, he agreed that it is too good to pigeonhole any longer. However, he did ask that only his first name to be used and he selected the name Howard for his student. So here is Leland's contribution:

I was a helicopter instructor at Ellyson Field in the middle Sixties as the Advanced flight training program was turning out scores and scores of Marine chopper pilots to feed the war in Vietnam. We worked long hours, early mornings into late nights. Rarely was a chopper on the deck. Everything was in the air.

I enjoyed instructing Howard, a highly motivated Marine Second Lieutenant. He was hell bent to get to Vietnam and lend a hand to his Marine infantry brothers as they waded into combat.

Howard and I were flying last launch one afternoon in an HO4S (Sikorsky, H-55), when he began to tell me about a nudist colony on the Eastern Shore of Mobile Bay. "About 20 miles north of the Gulf beaches," he said. "It's well hidden in a field of scrub pine trees, next to a pretty lake."

"How do you know?" I asked him.

"Word gets around, sir."

"Have you been there?"

"Well, I have to admit it."

"You crazy fool. Don't you know that's out of our flying area?"

Howard confessed that not only had he been there once, but four times. "You ought to see the white butts scatter. I'd come in low so they wouldn't hear my approach. Then I'd pop up over the row of planted pines and drop down into the clearing and pull in the power."

"I could bounce you from the program, Howard."

"I really wasn't dangerous, sir. I flew down there as the Navy was changing the color scheme of the planes and the side numbers were blotted out. No way they could ID me."

Then Howard approached me with, "Back you down, sir. Let's go down there today. Our side numbers are still painted over. All we have on today's training schedule is auto-rotations. I got 'em cold."

He was right. Howard could do auto-rotations like Sikorsky himself.

"Howard, I don't think it's a good idea. Both of us could lose all we've got."

"Sir, I'll never make mention of it again. Ever."

Howard was persistent and I don't know why, but I agreed.

"If I lose my wings, I'll kill you. You'd better get us back on time. You know how strict they are about being late in the chocks." Thankfully it was late afternoon and getting darker. I felt sure we couldn't be identified.

In minutes, Howard was flying low to the deck, fully out of view from anyone associated with Naval Aviation, I hoped. I just sat back and let him go. Why, I don't know. But I did know he would be a good helicopter pilot in Vietnam.

We traveled in a westerly direction, passed north of the outlying practice field at Robertsdale, raced past Daphne, hit the Eastern Shore and turned south. We dropped to below 100 feet and stayed a couple hundred yards offshore. Soon we passed Fairhope on the shore line. We flew over that big Grand Hotel and I still couldn't believe this was happening.

Farther south, the Eastern Shore took us in a southeasterly direction. Three miles south of US Highway 98, Howard turned to the east, dropped to maybe 40 feet and maintained his 90 knots.

"OK, sir. Few miles to go," he said. "Coming up in a couple minutes. Hold tight and keep your eyes open. You'll get the sight of your life, sir. Take over if you want to."

Now he was flat down on the deck as the line of planted pines appeared. He pulled up sharply and then dropped back down and slowed to maybe 50 knots. There they were. There must have been two dozen flour-white butts scrambling for the bushes. I really couldn't believe this was real. It was the craziest flight of my life. Howard turned back to the west and retraced our route.

"You keep the stick, Howard. I don't have the strength." With Howard flying this absolutely illegal route, I was too exhausted and wrung out to take control of the aircraft.

We actually made it to the chocks on time. When I debriefed Howard in the ready room, two other instructors were doing the same with their students. I didn't say a single word about the nudist colony that day, or just like he promised, not ever again. Howard got one of the most satisfactory flight reports that I had ever given any student. He deserved it. He was a superb rotary wing pilot.

I never returned to that site again. Howard graduated and I assume he got to Vietnam. If so, his Marines were very lucky.

Ray's Delimmas

Ray is a retired Naval Aviator, lives in California, and remembers a lot of stories about when he was in flight school — instructor and student.

Ray was a LTJG flight instructor just back from over a year's carrier duty in Korea in the middle fifties. Any time a student had a problem, the Navy was determined to get to the bottom of the story. No doubt, part of that reasoning meant trying to determine if procedures should be changed. On the other hand, maybe the student didn't have what it took. Or even, maybe the flight instructor wasn't up to par. Maybe he didn't learn all he should have in instructor school. In any case, the Navy wants all the facts.

Here are a collection of Ray's memories:

Ray's Number 1

Sometimes when things get bad, they can turn good. Ray Smith found that to be true as a student during a gunnery training mission in Advanced Flight Training in 1946.

I was a student on the perch at 6,000 feet and next to go on our dive-bomb training run. Like all of us, I was a little nervous. I kept raising the nose of my beautiful little F6F. There! Now! I was in position to turn in to the left. I nervously waited for the go command.

Just as I was given the go, I raised my nose too high. I stalled and fell almost into a hammerhead, one-turn spin, but quickly recovered and you would never guess what was in my sight. It was the target, dead center. I pickled and got a "bull's eye" on the radio.

The instructor came on the radio and said, "Now I want everybody to do it like Smith did."

Needless to tell, the immediate comments from the rest of the flight were, "Yea, like Smith did it, Oh sure, every day, etc." They were unmerciful.

Then back on the perch I was next to go and still nervous. Then I got the go and hardly even knew when I rolled over. I couldn't find the target until I was too low to adjust. God was with me. Suddenly it was dead in my sights. I pickled. Lo and behold, I got another bull's eye without an unintentional spin. I keyed my mike and said, "Nothing to it, Boys. Just pray!"

Ray's Number 2

I was an instructor coming back from breakfast with my students and saw it happen. A cadet, experienced in light planes but having trouble in heavier planes, was taking off in an F8F.

At about 400 feet his engine sputtered. He tried to turn back to the runway and at the same time stretch a glide. The result was a stall, split S, and an 8 ft hole.

It was a horrendous crash. My students started to run toward the crash. "Let's go see if we can help, sir," they yelled.

I looked at them, shook my head, and said, "Nothing to see, guys. You just saw the result. We're going to sit down right now and talk about stretching a glide."

Ray's Number 3

As an instructor once, I had a Cuban Midshipman A-stage student, George, having a hard time. But after finally getting him to fly with only two fingers on the stick and keeping his head in a windshield wiper motion on landing approaches, George soloed.

That night, he and his buddy showed up in his spiffy formal white uniform at my apartment with a bottle. Thomas, my VF-92 buddy, came over from his BOQ room down the hall and we finished George's bottle and one and a half of mine. I had to get someone to drive them back to the base.

The word got around. After that, any midshipman having any kind of problem tried to get me as their instructor. The word was that Ray's booze is good.

Ray's Number 4

In 1953 I was transferred to Pensacola after having served a year aboard a carrier roaming around off the coast of Korea. They assigned me to Saufley Field as a Primary Flight training instructor, teaching fledglings to fly.

I walked into my training squadron ready room one afternoon to see what good things they had for me that day. A note was tacked to the schedule board next to my name. "Ray, report to the Safety Office immediately."

My first thought was, "Now what?"

I walked topside and the door to the Safety Office was open. I came to attention and said, "Reporting as ordered, sir. This note was on the schedule board. What's the problem, Commander?"

"One of your solo students landed long up at Brewton this morning."

"Was he hurt, sir?"

"No."

"Was the plane damaged, sir?"

"No, he landed and ran off the end of the runway."

"Did he land long, sir?"

"Not too long. His nose gear was barely in the grass."

"Then it sounds like it was a good, maybe a passable landing."

"He landed kind long and ran off the runway. He called in, then walked up to the safety truck and reported it. They told him that a Saufley plane was already on the way."

"Any damage?"

"Well, no. But we had to send that plane and two instructors up to get him and return the plane."

"Commander, here's the way I see it. He didn't damage the plane. He radioed his situation, got out of the plane without moving it, and walked to the safety truck. He found that Saufley was aware and pilots were enroute."

"Well. I guess all that is true."

"Sir, I teach my students not to panic. Don't damage the plane. If you run off the runway, don't try to get back on. Report the incident and wait for help."

"Lieutenant, we had to send two instructors to get him."

"But, sir. He did exactly what I have instructed him. He's a good student. Maybe he needs a couple of additional warm-ups. The main thing I'm saying, sir, is that he followed my instructions."

"OK, Lieutenant. I'll give him two free warm-ups. He's sitting over behind the locker. Both of you get out of here."

I turned to see my student huddled, sitting, almost hidden from view, behind the open door. His head had been in his hands as he listened to us. Suddenly his very scared look slowly turned to relief. I put my hand on the student's shoulder and we walked out together.

Saved By Lightning

Tom Brannon finished college in 1970 and the Army reached out to him. "Army?" he said as he began looking in another direction. Read his story:

I served in the military. I was ordered to report for induction into the Army, but I escaped to the loving arms of the Marine Corps. Twenty-four years later I retired as a Marine Corps Lieutenant Colonel.

When I reported to the Corps, I requested Flight Training. I passed the flight physical, was commissioned into the Marines and sent to Basic School at Quantico, and from there to Flight School in Pensacola.

From Pensacola I was sent to advanced training at NAS Glynco near Brunswick, GA and soon went to VT-86. We trained in the mighty T-39, flying intercepts against A-4Cs.

One day over the Atlantic Ocean near Glynco, it came my turn to execute my five intercepts for that flight. Policies then in place stated that if two of the five runs were unsatisfactory, you got a down. My fourth run was an unsatisfactory and my fifth was not much better. As I recall, in order to be given a down, your intercept had to be "irrecoverable" when you were aft of the bogey's wing line.

The weather during that hop was atrocious – massive thunderstorms, lightening, driving rain, etc. Suddenly I was forward of the wing line of the A-4 and knew that I had blown the intercept.

Just then a bolt of lightening struck the radome of my T-39. The radome decomposed and most of it departed the aircraft. The cockpit depressurized. TACAN and most of the pressure driven instruments of the aircraft died along with the engine gauges and radios. I was knocked unconscious briefly.

I woke up just as the T-39 was deeply involved in a roll and it was not a roll-rated aircraft. I noticed that my radar was a mass of burned electrical wiring. It took a few seconds to quickly realize that my instructor could not possibly give me a "down" because the run was an "incomplete."

I was elated — not thrilled. Then I realized that my aircraft was in an extremely precarious situation at 24,000 feet. The flight flew back to Glynco and I blew the gear down for a NORDO (No Radio) landing.

I eventually got my wings. To this day I laugh at my stupidity then, but I am grateful for having been winged and for my service in six Marine F-4 squadrons.

Tom Brannon
LtCol, USMC Ret

My response to Tom:

Tom, I was a Marine Infantry officer in 2ndBn, 6thMar in Beirut - 1958-1959 before I applied for and was accepted for Flight School. Brian, a clever and creative PFC, and my radioman, was always reading some fancy literature book. We called him our platoon poet. He coined a phrase — ***to succeed is not to fail.*** *I believed Brian. I passed it around to a lot of officers and men who might have run across it. Not finding anyone who had ever heard this statement, we deemed it Brian's and hence named it "Brian's Call."*

Simplistic, but my guys liked it. My platoon sergeant found a local and friendly Lebanese lady in Beirut who made a guidon for us: a red background with those words, ***to succeed is not to fail,*** *in gold. In the bottom right corner of the guidon she embroidered the words, "Brian's Call" in small letters. I took that thought to*

Pensacola with me. I am sure my constant thoughts of "Brian's Call" pulled me through to graduation day.

You succeeded in a time of stress. And your story reminded me of that special guidon. That was a long time ago. Wish I still had it, but we gave it to Brian after we returned home.

Thanks for reminding me. Tom.

CHAPTER 13

Instructor Stories

No doubt exists in my mind that every instructor who ever set foot in the training halls of Naval Aviation was dedicated. You could read it on their faces, see it in their eyes, and hear it in their voices. It mattered not whether they were performing in ground school or flying positions, they were good. They wanted us to be.

From the author's memories and notes:

I remember ground school instructor, Lieutenant Brewster, who taught engineering at Saufley. He was positive we would learn how planes flew, where they flew, why they flew, and what caused them not to fly. He was good.

A week or so into the course, he gathered the twenty-eight of us (Some from Pre-flight were now gone) around him in a sort of first grade style circle. We moved in closely, sat on the floor, crossed our legs, and watched as he explained in detail his mock-up of an aircraft engine. He slowly followed the intake, compression, power, and exhaust strokes through its cycle, and explained each.

At the close of that day's session he had us copy this following poem on our pads. He said that God himself had authored it:

> ***If I don't buckle down and learn,***
> ***Then I shall surely crash and burn!***

Thank you, Mr. Brewster.

Andy Travers was a flight instructor at Saufley for three years beginning in 1957. In 1958 he arranged with his squadron commander to consider pairing instructors to student and maybe better students might come through the program. Here is Andy's story:

> *The most important job for the scheduling officers, especially in the early stages of Primary training, was to fit instructors to students as much as possible. Case in point, in A-Stage Primary at Saufley in 1959, our squadron commander, LT. Winters, placed a memo on the big board, a standing order to all students and instructors. He decreed that instructor-student changes could be made, without explanation, during the first three A-stage flights, A-1 through A-3. His theory was that if a miss-match occurs in initial assignment, the student should get a mulligan. Ample time would remain for the student to get ready for the A-12x (OK-for-solo check flight).*
>
> *Feedback from this new squadron policy showed that each of* ***two*** *instructor-student changes while I was there was actually made at the request of the instructor, not the student. That seemed strange to me. Why? The policy remained in effect as long as I was at Saufley. I did no further follow-up and it still remains a mystery.*

My Argus C-3

Jeff, a former flight instructor, thought at the time of this incident that he would be in serious trouble. As you'll see, he got by, although his flight student suffered a bit. Actually, it is pretty easy to slink out of trouble if no one knows you're in it. Jeff must have been a pretty nice guy. Read his story about his student named Ron:

I thoroughly enjoyed my assignment as a Radio Instruments instructor at NAS Whiting Field in 1961. Flying Radio Instruments training flights kept me up to date and well-practiced on my instrument flying skill. Students coming into this phase were good pilots. They had completed Primary, Basic I and II, and Basic Instrument Training. When they were under the hood, they possessed good headwork, were proficient in keeping the wings level and the plane on altitude as well as maintaining airspeed.

Radio Instruments added another dimension to the students' training as well as mine. Students learned to follow Air Traffic Control clearances to get under the hood, go from Point A to Point B. Upon arrival at Point B they would get an inflight clearance to Point C. Upon arrival at Point C they learned to get the plane home and on the ground as directed.

Most of my students had hobbies or interests other than flying. Coast Guard Ensign and flight officer Ron, loved flying, of course, but when not in the air, he enjoyed taking pictures with his brand new Argus C-3. He'd usually put his two loves together and nestle that little camera inside his flight jacket, ready for action. He used it at every chance.

It was first launch early on that morning in March of 1961. Ron had finished preflighting our assigned T-28 for his next Radio Instrument lesson and waited on me. As always, we took off from South Whiting Field and crossed Pensacola Beach at a thousand feet, then began our climb to altitude over the Gulf of Mexico. With Ron seated in the back seat, the bulky white hood completely blocked his view of the earth below.

While climbing to our work altitude for the day, 25,000 feet over the Gulf, I continued to spew practice clearances to my student. Once in our designated block of air space, we began working with other types of Air Traffic Control procedures.

This would be a double syllabus flight for two and a half hours: two lessons in one flight. After an hour of sweat and toil, I took control of the plane, turned back to the north, and allowed Ron to come out from under the hood for a rest lest he go stark raving mad. He always enjoyed looking at the scenery of the beautiful Gulf of Mexico five miles below and breathing in the thin, minus 20 degree air. On this day Ron pushed back his canopy and, although encumbered by the bulky oxygen mask and heavy gloves, snapped a few shots with his Argus C-3.

Suddenly, two Air Force F-104s appeared and snuggled, one on either side, matching the slower speed of our T-28. Startled, I quickly called Whiting Field for instructions and was given a tactical radio channel and a code on which to contact the Air Force aircraft. The F104 flight leader said that they had received an alert of an unidentified aircraft entering the ADIZ (Air Defense Identification Zone) from the south at 200 miles.

Within a few minutes, back and forth communications between the 104s, our T-28 and Whiting Approach Control revealed that **someone** *in our T-28 had failed to activate the navigations transponder when we departed Whiting. Radio signals sent from this device would have identified our T-28 as a Naval training aircraft from Whiting Field. Soon, the alert was canceled, except that I knew I was in for a reprimand. I was the instructor and I therefore had full responsibility for the flight. I would take the fall.*

"Sir," said Ron, "could we get those 104s to move in closer so I can get a shot?"

"Why not? My ass is mud now anyway." I contacted the 104 flight leader. Politely, he complied. The 104s followed Ron's requests to position themselves in several arrangements around our T-28.

Ron snapped a dozen or more shots. Getting braver by the minute, he leaned out in the slipstream a bit too far. The 220-knot air flow snatched that precious C-3 out of Ron's grasp, snapped the carry cord and shot it and his *undeveloped souvenir shots hurtling downward on a five mile journey of about 82 * seconds toward a thousand foot deep eternal and watery grave.*

Yes, I was admonished for failure to activate our transponder, but finally returned to the good graces of the Navy without a disciplinary letter. Ron finished flight school and became a pilot aboard a Coast Guard P3V Orion sub-hunter.

** In rough calculations, Ron's camera accelerated, drag not considered, at 32 feet per second per second, reaching terminal velocity in about 11 seconds, while plummeting 2275 feet. For the next 71 seconds, the little camera screamed toward the Gulf waters falling 22,725 feet at a constant terminal velocity of 352 feet per second toward its everlasting resting place.*

Bombs Away

Gunnery was the shortest phase in Basic. Many students felt it was the most exciting. Each flight consisted of an instructor pilot in his 1475-hp T-28 pulling a 15-foot long white target sleeve, and three to four gunnery students, each in his own T-28 equipped with two .50 caliber machine gun pods hung under the wings.

Typically gunnery flights were composed of a cluster of four student aviators who would take off from Whiting Field at Milton, Florida and climb west to pass over two very tall smokestacks at the factory near Cantonment, FL. There they would turn south to pass over the water tower on Pensacola Beach and begin their climb to altitude over the Gulf of Mexico.

Assigned altitudes would change often according to situations, but at the time this story occurred, the instructor for the flight would maneuver his T-28 to a remote pickup point, snag the target sleeve, climb while avoiding populated areas, and reach an altitude of 12,000 feet at least 20 miles south of the shoreline. By this time, the student flight had reached their perch at 18,000 feet, established contact with their instructor, and waited for the "prepare for attack" signal.

Former Naval Aviator Hayden had a nerve to rat on a Gunnery student from flight school in the early Sixties, and submitted this story. Hayden did not use his real name because he was sure Sterling would track him down and make him pay.

His name was Sterling. He was a Naval Aviation Student Officer, having received his commission via NROTC from the University of Missouri. Sterling was one of those on-the-edge, marginal flight students. He constantly remained on the verge of being dropped. His good points included being likable and jovial, in spite of appearing a bit superior.

Few of Sterling's peers had established a close friendship with him. A lieutenant JG, he had served on sea duty out of Norfolk as gunnery officer on an APA troop carrier. He had always wanted to fly and after a couple rejections, was finally accepted.

Sterling came close to being dropped from the program in Formation phase, possibly because of his snobbish attitude, and was just beginning Gunnery phase. Usually when students reached this point in the syllabus, they were considered to be downright good aviators and would probably complete Basic Stage, go on to Advanced, and receive their Wings of Gold. But most students who had trained alongside Sterling believed he had reached Gunnery not because of his off-and-on upbeat attitude, but because his grandfather had been an admiral in WWI and his dad died in WWII when the submarine he commanded was lost on a patrol off the Japanese coast.

On this flight in February of 1961, the weather was clear and cold. The students, one at a time, would dive from the 18,000-foot perch downward toward the target sleeve as the instructor maintained a flight level of 12,000 feet. They would try to bring the target into alignment by aiming through the Norden gun sight. Once stable, they would fire off several bursts, break and climb back to their perch and be ready for another run.

Sterling committed several blunders. He surely continued to grasp the "I ain't long for this world" handwriting on the wall. Several times over the past couple of weeks, he had mentioned to some of his associates that before being dropped from flight school, he intended to make his mark. He said his mark would be visible for years, high in the sky, and that everyone seeing it would remember him.

Apparently, today was to be Sterling's day. After all ammo was expended, the instructor, still towing the target sleeve, dismissed his flight, broke away and followed his safety path over the remote areas east of Milton, and circled west toward Whiting. The students, with Sterling flying number four well to the rear of the formation, headed north over the water tower on Pensacola Beach, then proceeded along their usual route toward the smokestacks at Cantonment.

The student flight descended to 1000 feet as they approached the smokestacks in preparation for entering NAS Whiting's landing pattern. Sterling, flying tail-end Charlie and out of site of the rest of the flight, dropped even further to 600 feet and zeroed in on his target — the smokestacks. At his best estimate of the proper time, Sterling opened his canopy, visually aimed for a second or two, and hurled a sack of flour from the cockpit toward the tall stacks. He keyed his mike and radioed, "I did it! Long live Sterling!" He wouldn't be sure yet of success, of course, because the planes would be a well past the stacks toward Whiting before the flour sack would hit. Sterling would have to wait to actually claim his victory.

The student flight landed and taxied toward the Gunnery parking area. Two Naval officers flanked by MPs gathered at the four planes as they shut down. They waited for the students to climb out. One of the officers ordered the students not to speak and they were herded into a flight operations office and ordered to stand at attention.

A Navy lieutenant commander began, "Who is the 'John Wayne' in this flight? Who dropped something over the factory at Cantonment?"

After a few seconds, Sterling raised his hand. "I did it, sir. I hope it worked. Did I hit the stacks?"

The lieutenant commander ranted for a minute or two before explaining that Sterling's aim was not perfect. It was a bit off. Instead of making his mark on the smokestack, the flour sack sailed into the employee parking lot and shattered the windshield of a late model Buick. The inside of the Buick ended up solid white and saturated with lily-white flour and looked as if it had been in a North Dakota blizzard.

The lieutenant commander's investigation determined that the other students played no part, and were released. That day Sterling became a member of the infamous Sundowners Club, a roster of students who have acted so waywardly that before sundown of the day of the offense, a review board, in short order, had convened and voted for expulsion. At sundown that day, the Training Command had one less flight student.

The Singing Flight Instructor

Royce, a former Naval Aviation Officer student from Tennessee, tells this story about one of his most memorable flight instructors:

In 1960, I had just completed Naval Aviation Basic Flight Training in Pensacola and checked into Advanced in Brownsville. I was hot to get into that T2J.

After a few days of concentrated ground school, ejection seat training, altitude chamber and several other events, I met my instructor, Lieutenant Lowry, in the ready room of Squadron 11. We seemed to "hit it off" as we walked to the flight line.

He began demonstrating the lengthy and detailed preflight check of our designated aircraft. We taxied out and took the active runway. Lowry very cleverly talked me through a flawless takeoff and in no more than three of four minutes we broke through the overcast at 30,000 feet over the Gulf of Mexico. I was proud to be flying this aircraft and, even at this early stage, felt fortunate to have this man as my instructor.

Lowry was good. He would explain a maneuver, demonstrate it, and turn the plane over to me. He would keep silent as I made mistakes. Then he would talk me through the maneuvers until I got them right. He helped me to ease into the art of jet flying.

One afternoon I had control of the aircraft, trying to emulate his smooth abilities. I began to hear a sound. Sounded like somebody singing softly over my headphones. I recognized the tune. It happened to be a country music song, "All My Exes Live in Texas." Occasionally the singing would stop and Lowry would take over and help me in my flight performance. Then the singing would resume.

This went on for the entire flight. During post-flight debriefing, Lowry gave me a well done and asked for my comments. I commented on the flight, but not about his singing. Back in the squadron ready room, I mentioned it to some students. They told me it was about Lowry's love life. He met wife number one in a local bar in town. They got married. A month or two later, she divorced him. They all thought she was just out to get a dependent's commissary and PX card. Probably so. They told me not to mention his singing, but just to sit back and enjoy.

I graduated, served my three years and got out in 1964. I went home to Birmingham, got my CPA from

Auburn, and went to work for an insurance company. Over the years I thought a lot about the lieutenant and what a sad story. In 1985, I decided I'd try to find him. I knew he was from Knoxville, so I started there.

Lowry is not too common a last name and I hoped it would narrow the search. I finally used some resources of my employer. Maybe this wasn't legal, but it was for a good cause and did no harm. A few days later, I found a full name match in a little mountain town a few miles from Knoxville.

I had to find Lowry. I talked to my wife. She was supportive. I took some vacation time and early one Saturday morning my family and I packed up and headed in the direction of Knoxville, then to the tiny town where I hoped to have a reunion with Lowry.

The local police were helpful after I showed them an aging Naval Aviation log book with Lowry's name as my instructor on several flights. I lodged my family into a 60-year old, immaculately clean motel and drove the ten mountainous miles toward a little cabin.

There it was, just as the cops said. I walked onto the porch, hoping the boards had not decayed as much as they looked. Suddenly I heard the sweet mountain sounds of a dulcimer in the background, and a soft voice singing, "One of my exes lives in Texas; the other left for old Saigon."

Chills ran over me. I knocked and heard a voice inviting me in. There was Lowry, his years showing more than they should. He was in a wheelchair on the back porch. His right hand cradled a .38 revolver.

"Lieutenant Lowry?"

"I made lieutenant commander," he answered, "It sure cost me a lot though. I'll be glued to this chair till the Lord takes me."

"I'm sorry, sir."

"I think I remember you. If I'm not mistaken, you were one of my students. Had to be before Viet Nam." He offered his hand.

Lowry began the story of the rest of his life. In 1968 his squadron was deployed to Southeast Asia. He would rotate between carrier duty and shore duty, one week at the time. While ashore near Saigon once, he met Choa, a Vietnamese girl. They were married in a Vietnamese civil ceremony and Lowry hid it from the U.S. military.

A few months later he was strafing concentrations of Viet Cong north of Saigon when a surface to air missile locked onto him. He pulled the face curtain but not soon enough. The SAM hit his wing just as he ejected. He lost both legs. A Jolly Green Giant Search and Rescue chopper found him the next day, barely alive and close to a river.

Lowry's war was over. He was shipped home immediately without seeing Choa. At Walter Reed, he recovered about as well as could be expected, so after his discharge he came home and bought this little cabin.

In April 1975, the U.S. pulled out of Vietnam. Thousands of Vietnamese relocated to the States. Amazingly, Choa made the list. She located Lowry and enrolled in the University of Knoxville, graduated, and became a teacher.

By 1983, Vietnam had relaxed its restrictions on certain former Vietnamese citizens who left. They invited Choa back to her country if she would agree to teach for five years. Lowry was also invited, but declined. When she left, both doubted they would never see each other again.

Lowry said he really loved Choa, just as he claimed to have loved his first wife. "But I'm content now to sit on my back porch, sip a Budweiser and sing my song." He began, "One of my exes lives in Texas; the other left for old Saigon.'" He didn't attempt to hide the tears. Nor did I.

He asked me to follow him as he powered his wheelchair along a well-traveled path that ended at a little clearing — his devotional area. Every day he spent at least a half hour in this spot in contact with his God.

After a long visit, I hugged him and shook his hand. I told him I'd be back again, and when I did, I'd bring my old log book and we'd relive good times. His face brightened.

Not long after, I went back to see Lieutenant Commander Lowry again. His physical condition had worsened. We looked over my logbook, visited his devotional area, and prayed. I asked if I could come back at Christmas. He was happy.

In December, a week before I planned to leave, I called the little police station to check on him. Lowry, the best flight instructor, and probably the best friend I ever had, had died. I am so glad I visited him those two times.

The Venus Fly Killer

Floyd, in submitting this story for the book, said it is amazing when he thinks back and remembers some of the leisure pursuits in which some of his flight students, and especially this one, indulged. Floyd was a Naval Officer and pulled two hitches as a flight instructor. He lives in retirement in La Jolla, CA now and sent in this story:

It was 1964 when I first met Wallace. He was one of the smartest students I had ever had. Wallace grew up on a small cotton farm in Mississippi, had few advantages, but somehow wrangled an NROTC scholarship and graduated from Ole Miss. His goal was to teach zoology at his alma mater one day. Any time he had a half hour before a flight, would get out his tablet and draw fish. He didn't have time to actually go fishing but he enjoyed drawing them. That's right, fish, all kinds of fish, the kind you catch if you throw a hook in the water and use the right kind of bait. His drawings were beautiful.

And then there was Corry, a young Ensign from New Jersey. Corry had had a much more comfortable upbringing than Wallace. He attended college in Indiana, also under an NROTC scholarship. He was thoroughly fascinated with the history of his black race. He would bring in different history books every day to study. Corry spent his wait-time collecting facts, arranging them for a future book, and planning for that day he would publish his book.

Other students involved themselves in more mundane activities such as pitching pennies to the line; paper, rocks, and scissors; flipping coins; even crazy eights cards. None of these events required much brainpower and that was probably the point. When students knew that their next flight could result in a down that might ultimately drop them from the flight program, no doubt exists that they would have concerns. And it was day after day after day.

I'm no psychologist, but I have an idea that these strange leisure pursuits might be because of the constant ultra-stress of flight school and how it plays differently with some kids.

Another student of mine, Darrell, developed a love of killing flies, house flies; the strangest student pursuit I ever ran into.

The huge sliding doors to the hangars at Whiting Field that housed the training squadrons remained open in cold weather or hot, rainy or dry. Even stray dogs and cats from distant places would detect the aroma of the hot dogs and hamburgers cooking on the grills and come a-begging.

In the summer time, flies were the worst. The Navy sprayed around the hangars often. That might resolve the problem for a while but they'd soon be back.

My student, Darrell, came up with a novel and creative idea. He'd get a rubber band and, in a motion similar to someone shooting a young boy's sling shot, would ease up close to a fly, pull the rubber band back, wait, and snap. Darrell became rather adept at slaying flies. During times of good hunting he would display a sizable number of kills on top of a sheet of notebook paper in the ready room. He claimed to be able to predict, using Navy weather information, when flies were most likely to collect.

Darrell and I were flying over the Gulf one day, learning to execute the Immelmann aerial maneuver. He suddenly said, "Sir, can you take the plane. There's a fly up here bothering me."

What else should I do? He was a good student and always tried hard. I clicked the mike, "OK, Deadeye. I got the aircraft."

Although his body and the seat blocked most of my view, I saw him slowly lean forward. Then he became deathly still. Suddenly he threw his hands in the air and said over the mike, "Got him, sir. I'm ready to go."

"OK, Deadeye. You got the plane."

When we landed, he took that fly into the squadron area and wrote a full account of the "battle" on a sheet of paper. It remained on display for a few days and was published in the Whiting Tower, with photo, of course.

Duncan

The Navy does a good job of selecting Naval Aviators to be trained as instructors. Once in a while they will, by design or by chance, come up with a superb teacher. This is proven in the following story. Read Marvin's remembrance of a man, an instructor, and a friend, Lieutenant Commander Duncan who called himself Dunk:

We never did know whether he spelled his nickname Dunk or Dunc. He said it didn't matter. We did know that anytime he had a minute and we had any kind of problem, he was available. We didn't know any other instructor, especially a LTCDR instructor, that would stroll around the ready room, speaking with students doing paperwork, maybe waiting to fly, filling out a debrief report, or maybe

studying for a math exam. If you had a dilemma, or just a simple question, just say, "Hey, Dunk. Gotta minute?"

He always did. Right then, or for something that might take longer, he would set up a time. He was tough during a flight, but easy to talk to on the ground. He probably reminded us of our fathers: tough, demanding, but kind.

This story goes back to February of 1959 at Whiting Field. My flying partner, Bud, and I had just begun two-plane formation stage in the T-28. Bud was erratic, especially on cross-unders, and our instructor in the chase plane kept correcting him. A couple of times chase told Bud that the object of a cross-under was to ***cross under*** *my tail section; not to* ***fly through*** *it. It was scary.*

The instructor gave Bud a flat unsatisfactory that day and recommended a re-check. We'd repeat the flight if his re-check warranted. Our debrief ended about the time I got a phone message saying my wife was in the hospital at NAS, about to deliver our baby.

The squadron CO told one of my buddies to drive tandem with me to the hospital. I followed him slowly, thankfully. I was nervous. I don't think I could have safely made the hour trip by myself. Without a doubt, part of my anxiety was replaying that flight and thinking what might have happened if Bud had rammed my tail section.

When we arrived, wife and baby were fine. Several students' wives from my class were already there and showing support. It's the Navy way. I called the squadron. It was Friday and the CO asked if I could take Monday off and return on Tuesday. I agreed. A three-day respite was nice but it gave plenty of time to think. I began to think baby, cross-unders, and Bud, and over again. I had never felt like this.

I returned Tuesday to a morning flight, again with Bud as my flying partner. I was a rough, but safe. Upon our instructor's command, I took the lead with Bud on my wing. After a few simple maneuvers, chase instructed us to fly a pattern where I would fly 500 yards dead ahead of Bud. I would break and execute a 180 degree turn and Bud would break and join up on me and execute a cross-under. Then we switched so that Bud would lead and break, and I would join up and execute a cross-under.

It seemed to go fine. We executed properly. Chase called for one more run with me in the lead and Bud joining with a cross-under. But lo and behold, this time instead of a cross-under to my starboard side, Bud's plane blasted by me at an extraordinary rate of speed, missing my tail by inches. He sailed far out, off my starboard.

Chase began yelling obscenities. He yelled to Bud, to no avail. Bud finally came back and chase ordered him to take up a heading immediately for Whiting Field. I was told to join up on Bud but to keep a wide berth. The instructor didn't say a word all the way. I didn't think I had a burden in this incident, but I knew Bud had a bunch.

Chase's ready room debriefing brutalized Bud. In his attempt to join up, he did not stabilize off my port. He didn't even stop. He sailed directly behind me at high speed, his prop missing my tail section no more than a foot. The instructor ranted that he could imagine two dead bodies floating in the Gulf.

He downed Bud with a recommendation that the flight board drop him. I got credit for the flight, but for the entire day my mind clouded with anxious thoughts about cross-unders, babies, and Bud. Bud wouldn't be back, but someone, maybe worse, might replace him. At home, my mind clouded as I stood and looked into the crib at the baby and recalled the instructor's admonitions.

The next morning as I drove to Whiting, my stomach tightened, my breathing accelerated, and I wore a booming headache. I tried to recover. I couldn't. I had no idea what my morning flight would be like. I said, or actually lied, to my instructor that I must have eaten something that disagreed with me. I was excused from the schedule. I'm sure he knew.

The next morning was the same. I arrived early, found Dunk, and we went into a private debriefing room for several minutes. He told me to sit tight. In a few minutes he said he was taking me on as his student. I would be off the schedule until further notice. That worried me. I was to go home, rest, and report directly to him the next morning at 0730.

After a horrible night, one that I saw myself dropped from the program, I met Dunk. He told me to make no mention of anything I would be doing. I would report to a specific base doctor at 0900 this morning.

The doctor was a shrink. I spent one half-hour session with him that day. The next day I returned for another half hour, after which, the shrink instructed me to go home and report to Dunk at 0730 the following morning.

At 0725 the next day I walked into the ready room. Dunk said, "I got us a plane. You ready to fly?"

Before I could think, Dunk and I were in the air. We flew places I had never been. He demonstrated dive bombing techniques I'd be doing in Advanced. We flew low and fast, doing strafing runs on the eastern part of Santa Rosa Island. We flat-hatted the beaches from Panama City, along the Gulf, to New Orleans, even "dragged" Lake Ponchartrain at 30 feet of altitude. Dunk would not take the controls. I relaxed. He wrote up the flight as a warm-up.

The next morning I was in the air, with a new flying partner, on the regular 0800 launch on a syllabus formation flight with Dunk as his student.

The Navy way for sure!

CHAPTER 14

Instructor Bios

Author's Notes on this Chapter: As I began to write this book, I made a request from Jay Cope, PAO Chief and Lori Aprilliano, PAO Officer at NAS Whiting Field. I wanted my research to include interviews with a couple of top-notch flight instructors and some equally first-rate students. I asked for ample interview time to get biographical sketches, personal glimpses, and, of course, any interesting stories that they might choose to present.

Jay and Lori assisted in the selection of two instructors with whom I became thoroughly impressed: Navy LTCDR Lena Buettner, and Marine Captain Chris Phillips. I had hoped to include male, female, Marine, and Navy. These two individuals combined fit the bill.

After completing my interviews with the two instructors, I knew that I had scored a hit. An unanticipated wealth of information had poured out. And further, I made some good friends. In the old days, if either of those had been assigned to me when I was a flight student I would have rejoiced. Furthermore, if either had been assigned to me instruct, I would likewise have rejoiced.

Naval Aviators reporting in as instructors are sent to Flight Instructor Indoctrination Training and receive rigid education in the ***art*** **and** ***science*** of teaching students to fly. Not every candidate will be successful in passing this course of instruction. Should each? With the stretch of your imagination, consider whether an unqualified instructor might possibly ruin a student who otherwise could succeed with a more qualified teacher.

A *qualified* student is a precious commodity and a very important cog in this huge wheel called Naval Aviation Flight Training. Few young men and women meet the entry requirements. If you recall on the first page of the Preface of this book, this author quoted from a commander's welcoming speech to his incoming Preflight class in 1959 as to how few applicants may qualify for selection.

The flight instructor is just an important cog in the wheel. In this author's over a year and a half in flight school, I met and adapted to every style of instructor imaginable. Most wanted us to succeed and would walk an extra mile if we were having difficulty. We had the utmost respect for them. Very few instructors were those I never wanted to see or even think about again.

My two instructor friends, Lena and Chris, wrote very well, which is a relief to this author because I work very hard to edit documents (strictly for readability and not substance) pouring in from so many non-writer types. The documents submitted by Lena and Chris remain fundamentally as submitted.

It is interesting to say that as I interviewed Lena and Chris, personally and by email, the respect that I always felt toward flight instructors during my flight training was still there, prompting me to render a "yes, sir" or "no, ma'am" several times. And I am double-plus their ages.

Lieutenant Commander Lena Buettner, USN Flight instructor in HT-8, HT-18, and TW-S NAS Whiting Field

One of this author's most enjoyable visits to NAS Whiting Field in preparation for writing this book was the opportunity to interview instructors and senior students. I was introduced to LCDR Lena Buettner, USN, flight instructor in the TH57 helicopter. Lena's stories:

I was an ensign flight student. My instructor, Colin, utilized three methods to teach me the basics of flight training in the T-34 Turbo Mentor. First, he used the ***threat method****. On my first flight in primary flight school, I inadvertently uncurled my grip from the stick and braced my hand against the side of the cockpit as the aircraft hurled down the runway. I didn't know that I had committed this outrageous crime until we had landed and we were walking to the line shack. My instructor turned to me, told me of my error, and proceeded to implement the threat method. "If you ever take your hand off the stick during takeoff again, I will never fly with you again." His threat was quite effective as I kept my hand firmly glued to the controls on every flight after that.*

The ***replacement method*** *was Colin's second method. I was strapping myself into my seat before one flight when he appeared beside me. "Lena, have you ever ridden a horse before?"*

"Once, sir," I replied, not knowing why he was making polite conversation as we're about to embark on a flight.

"Did the horse do what you wanted it to do?" he continued.

"Well no, sir," I said, still wondering what this was leading to.

"You had to make it do what you wanted, right?"

"Yes, sir."

"Well this plane is your horse," Cowboy Colin noted. "You may give this beast a few Whoa boys and Giddyups throughout the flight, and find that this does not work, as altitude, airspeed, and pattern checkpoints still rear out of control. Unless the T-34 replaces its parachutes for saddles, this method must be best left in the stable.

A final method utilized by my instructor was the ***supernatural method****. On one flight I'd perform okay (I won't go so far as to say 'good') and then the next flight would be horrible. Colin was convinced I had an evil twin sister. What was his idea to prohibit this entity* *from coming on board? One of his suggestions was for me to hold nightly séances to ward off the evil sister. This method I did not try, so I can't attest to the effectiveness. Perhaps it would have been a good choice to use if I ever had ENS Psychic Cleo as my instructor. Colin's best bet was to stick with the threat method for future success.*

As a flight student, I have felt the anxiety of waiting for the light bulb to turn on, the frustration of being cancelled for the third day in a row due to bad weather, and the disappointment of failing a flight.

Colin's ability to teach worked. I experienced the delight of learning to hover for the first time, to auto-rotate full to the deck, the sense of accomplishment of earning the right to solo, and the pure joy of my family pinning on my Wings of Gold.

Lena completed flight school in 2002 and joined the Fleet. In 2006, she was ordered back to NAS Whiting as an instructor. In the following story, Lena refers to CRM (Crew Resource Management). This concept combines the following seven *learned* skills:

Decision Making
Adaptability/Flexibility
Mission Analysis
Communication
Leadership
Assertiveness
Situational Awareness

In Naval Aviation, these seven skills work individually and jointly for safe and efficient flight operations. If this concept were to be instilled into every grade school child imagine how much better and efficiently the world would operate.

CRM During an Emergency

As an instructor, my student and I were conducting weekend operations based out of Pensacola Regional Airport in 2007. We had just taken off runway 26 at night for a Night Vision Goggle (NVG) training flight. I was the flying pilot on the controls and my student was the non-flying pilot, navigating with the use of a chart.

Approximately one to two minutes after takeoff, the entire cockpit filled with smoke and fumes. The air became was so thick and odorous that we were having trouble breathing and talking. I immediately turned back towards the runway, reached up to turn off the air conditioning, and declared an emergency with the Pensacola Tower controller.

Since I had taken care of securing the air conditioning, the first step of the Smoke and Fume Elimination emergency procedure, my student began to take care of the remaining steps of the procedure. During this time, the NVGs felt very constricting and claustrophobic on our faces in conjunction with the heavy smoke and fumes. Our tendency was to reach up and push the goggles away, but if possible aircrews should remain goggled during an emergency. The last thing that an aircrew wants to experience during an emergency is a different sight picture than the one they're used to. We knew it was best to leave the goggles on.

The tower controller cleared us to land anywhere on the airfield and told us she was sending the fire trucks. Then she proceeded to determine the nature of our emergency and whether we needed assistance. I could not answer her as I was trying to safely land the aircraft with a now-tailwind.

Additionally, several construction cones, marking a closed runway, were brightly flashing causing the NVGs to degain, making it difficult to see properly. Tower continued to communicate with us.

My student asked if he should respond. I told him, "Aviate, navigate, communicate. I am aviating and if you're able, you communicate." He was able to tell her to hold her questions for the moment. I landed the damaged aircraft and was greeted by a fanfare of fire trucks on the runway. We shut down without further incident. The next week, I learned from maintenance that the air conditioning condenser had experienced an electrical fire, causing the smoke and fumes.

As an aircrew working together, we completed this emergency exactly as briefed and exactly as we teach our students. My student immediately went through his steps of the procedure without prompting from me. Also, as the non-flying pilot, he backed me up as I safely flew the aircraft. This story is a training command success story all-around.

Comical non-effective CRM during an emergency

In Naval flight school, emergency training is an important part of students becoming competent aviators. Not only is the knowledge of emergency procedures (EP) crucial in case students experience an actual emergency on a solo flight, but the repetition and high op-tempo of EP training creates the skill set of the student being able to accomplish many tasks simultaneously in the cockpit. When an aircraft experiences an actual emergency, the student is really able to learn these skills first-hand.

During a formation training flight, my aircraft experienced a generator failure. My student and I completed the emergency procedures and notified the other aircraft of our problem. We were at a Navy Outlying Field at the time, approximately 8 miles from our home base. The procedures for a generator failure called for a practical precautionary landing as soon as possible. During daylight and good weather conditions, we should fly the aircraft to home base for the landing.

The other aircraft assumes the lead position in the formation and flies us home. In completing the steps of the procedures, we secure all unnecessary electrical equipment, including the VHF radio. We leave our UHF radio on so we may hear communications with Whiting Tower and Ground Control.

In this case, unbeknownst to us, the lead aircraft attempted to communicate with us on the VHF radio while we were coming in for the landing. Since we had secured this radio, we did not hear their communication, so we could not reply. At this point, the instructor in the lead aircraft assumed that we had experienced a total electrical failure and he declared an emergency with Whiting Tower, asking for immediate handling. Whiting Tower ordered all other aircraft to cease movement and they cleared the airfield for us. My student and I heard this happening since we still had use of our UHF radio. This wasn't the case at all but there wasn't anything we could do at that point to rectify the situation. So we enjoyed the status of immediate handling and landed without incident.

After both of our aircraft had shut down, the two instructors and the two students debriefed the event and we all learned a lesson in non-effective CRM that day. Even though the incident was not harmful to anyone, in retrospect, I should have told the other aircraft that I was securing my VHF radio and he would assume that I had done this since it is a step in the procedures.

The Hidden Perks of Being a Flight Student

As a flight instructor, I've felt the impatience of waiting for the light to turn on with students, the discomfort of the "Helo Hunchback" resulting from spending hours and hours in a small cockpit, and the sorrow of failing a student on a bad flight. More importantly, I've experienced the elation of teaching a student to hover for the first time (to hover is divine), the pleasure of flying students up to my home state of Wisconsin for an air show, and the satisfaction of watching my onwings receive their Wings of Gold.

During a low-level Search and Rescue (SAR) training flight in Florida's East Bay, my student and I came across a couple of alligators floating in the water. My student immediately turned ashen and fixated on those monsters. I asked him if he was nauseated from the low altitudes over water. He said no, but told me that he routinely took his children water-skiing in this area, not knowing that it was infested with gators. No doubt he changed locations the next time he took his children out for a weekend of water-skiing.

Captain Chris Phillips, USMC Flight instructor in HT-8, HT-18, and TW-S, NAS Whiting Field

Captain Christopher "Barbie" Phillips, USMC, originally from Swansboro, North Carolina now calls San Diego, California home. He grew up dreaming of becoming a military flyer. He chased air shows and read everything he could get his hands on about airplanes. This love of flying machines once made him think he would grow up and join the Air Force. Besides, that's where the pilots were. But growing up in a Marine Corps dominated area he learned that those crazy, short haired, and extraordinary Marines had pilots too. It seems that he couldn't resist the challenge of being a Marine so in 1998 he signed on the dotted line to join the *gun club*...or at least to audition.

Chris wanted to become an officer, so he entered the Marine Corps' Platoon Leader's Course while in his freshman year at the University of North Carolina at Chapel Hill. In August 1999 he found himself undergoing the trials at Officer Candidate School, just as many great Marines have done before him. It was a tough summer but he graduated, proving to himself that he had what it takes to be one of the Few, the Proud, the Marines.

In 2002 Chris' title changed to 2ndLt Christopher Phillips USMC. The next step sent him to The Basic School to learn what it takes to command a platoon in combat. Following graduation from TBS he packed up his Jeep Wrangler and watched Washington DC fade in the distance and set his sights set on the infamous Pensacola, Florida, the cradle of Naval Aviation.

After the weeks of Aviation Preflight Indoctrination (API), Chris found himself with orders to Corpus Christi, Texas to fly mighty T-34C mentor in primary flight training. Bad weather dragged out his stay to 10 months, but allowed him to meet a lot of Marine Corps helicopter pilots and to become engrossed with their tales of flying low and getting dirty. They were a great bunch of men. Not knowing previously what he wanted to fly in the Marine Corps, his mind was made up. Chris now put helicopters at the top of the list. Choppers were soon to be the focus of Marine Aviation.

After primary, Chris was awarded his choice so he packed the Jeep again and headed to Pensacola to learn how to hover those machines called helicopters. He checked into HT-18 for advanced flight training and like all new students immediately started asking who are the instructors everyone wanted and who were the ones to avoid. Chris and his familiarization partner found themselves with an instructor all said to look out for. The Navy LT was a hardcore and serious. He made their lives hell, but it was worth it. They toughed it out and became two great helicopter pilots.

During helicopter training Chris hoped that one day he'd fly the biggest, fastest, and most powerful helicopter in the United States arsenal, the mighty CH-53E Super Stallion. With the influence of one of his instructors, a Navy MH-53 driver, pushed him to win his wings and lock himself inside the cockpit of "Big Iron."

They flew several training flights together and had a lot of laughs. On one such flight, Chris recalls that they were scheduled to fly to downtown Mobile, AL for two training flights. The first leg went from South Whiting field to Mobile with no problems. While on the deck, a massive storm grounded them so their CO and OPS-O decided for them to spend the night in Mobile and then fly home the next day.

With such permission plus a rental car, and no clothes for this unexpected overnight trip, they immediately headed to the nearest watering hole like any good aviator would do. They bounced from bar to bar impressing the local females. Chris still remembers starting the night off as just a lonely student copilot and ending the adventure as an astronaut on a training mission. This was quite a feat in only one night in downtown Mobile. They managed to collect themselves by the next morning, climb into that helicopter, and fly it home satisfied with all the fun that was collected and the white lies that were spun.

Chris finished at Whiting field and pinned on his wings in June of 2005 with his dream orders to head off to fly the monster CH-53E out of San Diego, California.

In December of 2005, he joined his tactical squadron and shortly thereafter, checked in to HMM-165 to become part of the 15th MEU. After a few months of work-ups off the California coast he departed San Diego on his first deployment. By Thanksgiving their global hopping ended and he found himself in Al Asad, Iraq flying combat missions in support of the Global War on Terror. During that deployment they lost a few Marines, were extended a number of times, but eventually made it back to the ship to finish out their deployment.

It was a long deployment but finally by 2007 Chris was coming home from his first deployment. Typically, guys would get more time off but in a few months after coming home, but he volunteered to go back out. This time he was with the 31st MEU flying out of Okinawa, Japan. During this deployment he flew successful missions in Southwest Asia and returned home.

Still with the 31st MEU, Chris deployed again and during that final 31st MEU flew humanitarian missions to help the survivors of the Padang, Indonesia earthquake. Most Marines sign up to destroy the enemy, but flying those helping hands missions provided a sense of accomplishment as well.

Five years in a tactical squadron, including combat missions in Iraq, providing humanitarian aid in Southwest Asia, and even a flying role for Hollywood in the movie Transformers II, Chris earned just about every qualification a Marine helicopter pilot could obtain. He was ordered back to Pensacola to teach the future of aviation.

Chris taught fledgling pilots for a year and he says he is truly impressed with most of the students. Naturally, some are not quite there yet, or struggle for something, but that is why he and his other instructors are there.

Sometimes the fruits of labors are demonstrated. Chris recently had a former student, after his winging ceremony, come to say that he had selected to follow Chris' footsteps and fly the mighty CH-53E because he had made such an impression on him. Chris immediately recalled the Navy LT instructor who had made such a good impression on him that he went into CH-53s. Professionalism builds structures algebraically with no limit as to the good it will do.

Chris still loves flying and continues to spend weekends flying civilian airplanes on the side and getting qualifications there. Chris' ending remarks to me: "I guess I am still that little boy who is in love with flying and airplanes and now I am sharing my passion with the future of the Navy and Marine Corps."

It's a tough job living the life of a flight instructor. Not every Naval Aviator wants to be handed a set of change of duty station orders and find that he must now leave an exciting job in the Fleet and is headed for Naval Flight Training. Their duty now will be to trail a batch of students; this time to teach, not to learn. Some might be tempted to run for the hills.

Every instructor and student that ever was connected to flight school will recall great moments, some greater than others, when everything clicked. Likewise, every Naval Aviator and student ever connected with flight school will recall not-so-great moments; moments when nothing clicked. That's life.

Instructors sent in the stories for this chapter. They were there and saw their students as they tried to hide their errors. These students needn't have done that. It happens to everyone at one time or another.

CHAPTER 15

Where Do We Go From Here

What a great day! We made it. This author received lots of stories about graduation day. Why not? That day was the grand finale of up to two years of challenging work and study in addition to what seemed like a lifetime's worth of stress. Few people experience stress such as when an unintended blunder could have such grave consequences. The pages of this chapter speak collectively for Naval Aviators over the past hundred years in describing what we remember. The first is from the author:

Graduation Day

For almost two years we struggled, studied, sweated, and yes, prayed. Few of us believed that God would actually intercede on an individual basis for such a minimal cause as earning an earthly set of wings. He had more important things to do. Maybe prayer helped our inner feelings.

We newly-winged pilots were easy to spot after graduation ceremonies. We'd hurry down the street to the

PX wearing our newly acquired Navy Wings. A tradition had evolved so that after each graduation ceremony day, the PX would section off and decorate, in a Naval Air motif, a few tables in a prominent location. They would invite us to come and celebrate with free coffee and doughnuts. Of course our spouses, those magnificent creatures who supported us through good times and bad, were invited, as well as other family members present. We joked that over the couple of years, with the lunches and snacks, haircuts, and uniform pieces we had purchased, the PX could afford it. Yes, we had paid in full for that coffee, but it was a grand salute.

Eight of us graduated that day. We were easily spotted, dressed in Class A uniform, standing or sitting in small groups, sipping that coffee, eating those doughnuts, and bidding our friends goodbye. At least once each minute or so our hands would find their way to the left breast pocket of our uniform jackets and gently touch the golden wings, just to be sure they were real. We could hardly believe it.

We'd never see some of those friends again. We'd miss ones we comforted in their toughest times and those who reassured us in our periods of strife. Today, I think of some of my closest friends: Gus, Sig, Tom, Jerry, Larry, Bill, Donald, Gale, Dave and many more. Why can't I remember the rest? So many years have passed. Memories tend to fail. I hope they know that I haven't really forgotten. It is a brotherhood of the very strongest eminence. Few outsiders would understand.

Newly designated Naval Aviation pilots depart from the Naval Air Training Command. It's over. We head for new venues and new responsibilities. But training does not cease. It has barely begun. New planes, more powerful than the ones left behind, are to be studied and mastered. New procedures must be learned.

Greater responsibilities will be assumed. For the past year and a half we have strived toward only one goal, that of being good enough to earn those Wings of Gold. Every ounce of energy has gone into this goal. Now we must study and work to be good enough to keep those wings and to protect the country that has given us this opportunity. Now we will take on an additional dimension. Every squadron officer will have at least one secondary, non-flying duty. Young officers will be assigned to some major section of a squadron: administrative, intelligence, operations, engineering, or logistics.

Over the years, the method of travel has changed. As this book is written, air is the most common conveyance for officers who don't own an automobile. For those who do, we will travel by what is termed private-owned conveyance.

Prior to the Fifties, and especially during the years of World War I and II, trains, officially termed by the Navy as rail, were not only convenient, they were the choice decision. They extended to every nook and cranny of the nation. Trains were comfortable and kept to a rigid schedule as they continued their journeys over flatlands, mountains, deserts, and coastal areas, passing through the smallest of small towns and the largest of cities. This reach allowed us the opportunity to hop a train after graduation, head to see Mom and Dad for a few days, and then continue to our first duty station to meet a deadline.

During the WWII years, it became the Navy's mission to handle most of the air support duties in the Pacific area. The Army Air Corps with its thousands upon thousands of B-17s, B-24s, B-25s, and finally the mighty B-29s, escorted by long range Army Air Corps fighters at high altitude, were creating havoc in the European Theater. Fighting the Japanese required a different approach. Thank God and Eugene Ely, who on 14 November 1914 performed the first successful carrier landing, a highly successful landing on the rough flight-deck-equipped USS Birmingham, the Navy had seen the future and had answered the call, making carriers a major part of Naval Air.

Lights, Ladies, and Liquor

This story is submitted by Ranny, part of the *Greatest Generation*:

I enjoyed flight school. I don't mean it was easy. It wasn't, but it was what I wanted. As graduation time grew nearer, I realized the war was raging and that soon I would be in a cockpit facing some Nip whose greatest desire was to kill me.

But as a new ensign, I knew that when I got to the West Coast, I would go through more training before I headed to combat, so I'd try to have a little fun first. Instructors used to talk to us about all the 'sweet things' in the bars near the Navy bases. That sounded interesting.

When I graduated in 1944, my orders instructed me to proceed to Moffett Field in San Francisco. My instructor whacked me on the back and told me how much fun I had ahead of me. He said San Francisco was the land of "Lights, Ladies, and Liquor."

It sort of scared me. When I got home for my five days leave, I told my folks where I was headed but not what my instructor said. It was soon time to catch the train and head toward San Francisco. All my family came down to the station — as did my girl friend. I tried to forget what my instructor had said. But who could?

Days later the train pulled into San Francisco. I got off and had no idea which direction to turn. Then I saw a couple of Navy seamen waving signs asking all military going to Moffett to gather at their bus for a ride. I felt better as I boarded their bus.

As we motored across town toward Moffett, I started looking around for the Lights, the Ladies, and the Liquor. I saw plenty of lights. I saw a few places where I could get a beer. I didn't see all the ladies that my instructor said would be chasing me down.

I checked into my BOQ and I began to miss my family and my girl friend very much. I thought about the day I had left them when I had lights, ladies and liquor on my mind and I should have been saying goodbye. I sat down and wrote a long letter to my family and my girl thinking maybe this would help my feelings.

Contaminated Logbook

In the early 1960s, anticipating a new kind of war, many students, especially Marine students, who were already selected for the jet pipeline were frustrated when the Corps re-directed them to the rotary wing pipeline. Additionally, the Corps began to pull a huge number of fixed wing pilots from the Fleet and send them to Ellyson Field in Pensacola for transition to rotary wing. As a rule, they were not happy about this. One of the most common gripes among these transition students was that they detested the idea of having their pilot logbooks "contaminated" with helicopter flight time.

In 1962, this author was stationed once at MCAF Santa Ana, California in HMH-462, flying the Marine Corps HR2S (Sikorsky S-56). My next-door neighbor on Inchon Place, Carl, was attached to an El Toro Fleet training squadron, flying the F2H Banshee. He joked with me often about my transition to rotary, so I invited Carl to set aside time from his duties one morning to experience a rotary wing flight. We set the date and my crew and I flew over to the MCAS El Toro transient aircraft parking area. Carl walked toward the 15-½ ton rotorcraft with its five blades spinning 20 feet above the deck. As we lifted off at 0815 hours, the crew strapped Carl in and fitted him with a command helmet. We headed for the RAL (rough area landing) area, those mountaintop-landing sites, sitting 2000 feet or more above the terrain below, in the Horno Ridge Mountains, just east of El Toro. These fragile landing sites allowed helicopter pilots plenty of precision landing practice for flying into rough terrain.

We worked the RALs for a while then headed south toward the coordinates to meet a USMC artillery battalion at 0930 hours in the hills of Camp Pendleton for ninety minutes of assistance in the training of their new personnel. Our mission this morning was to work with the artillerymen to practice hooking up their 105 mm artillery pieces onto the sling under our aircraft. We'd approach station A and the handlers would attach the artillery piece to the hook. We'd lift off and fly to station B and let down so the handlers could release it. When all of the six guns were moved, we'd bring them back. We'd practice this over and over.

At 1100 hours we left the artillery site and flew westward to the Pacific Ocean shoreline. There we turned north, flying 200 feet offshore, and headed home. The crew chief asked if we could leave Carl with a pleasant memory of his flight. Permission granted, they seated him in front of an open port on the starboard side facing the beach, and handed him a pair of powerful binoculars. To make it even more interesting, we moved in very close to the shoreline, dropped to less than fifty feet of altitude, and slowed to about 60 knots of airspeed. A plentiful number of beauties sunbathed that day and all seemed to be waving at us. Carl talked about that trip often.

A year later we were transferred but received a letter from Carl and Ann. Carl had received orders to Ellyson Field in Pensacola and would soon have a contaminated logbook.

CHAPTER 16

Miscellaneous & Trivia

HIGH FLIGHT

Oh, I have slipped the surly bonds of earth
and danced the skies on laughter-silvered wings.
Sunward I've climbed, and joined the
tumbling mirth, of sun-split clouds—
and done a hundred things of which
most have never dreamed—
wheeled and soared and swung high in sunlit silence.
Hov'ring there I've chased the shouting wind along
and flung my eager craft through footless halls of air.

Up, up the long, delirious, burning blue
I've topped the windswept heights with easy grace,
where never lark, or even eagle flew.
And, while with silent, lifting mind I've trod
the high untresspassed sanctity of space,
put out my hand, and touched the face of God.

John Gillespie Magee, Jr., a nineteen-year old American pilot was flying with the Royal Canadian Air Force during WWII. On 11 December 1941, near Scopwick, England, Magee was flying a Spitfire on his last training mission before deployment into a combat squadron. He died when he collided with a fellow plane in his training group.

Channel 33

On Thursday, 12 May 2011, this author was driving west on I-10 from Tallahassee, Florida to NAS Whiting Field, Milton, for another interview marathon in connection with this book. I would be speaking with several Naval Aviation students and instructors and touring the magnificent new flight simulators. I had left home, just north of Tallahassee, very early in the morning in spite of the intermittently heavy rain.

Dawn broke and rain continued to fall. Starved for breakfast, I pulled into the parking area of McDonalds at Crestview, Florida to get a good old sausage biscuit. I parked and noticed an older lady, under the raised hood of her automobile. With a heavy downpour damaging her efforts at this moment, she was holding a raincoat over her head while she also tinkered. Three young tots, I learned later were grandchildren, sat inside the car out of the rain. I stepped over.

"Can I help you, Ma'am," I asked.

"I got to get Albert to fix this car, mister."

I pulled out my cell phone. "What's his number?"

"He ain't got no phone. Only one is at home."

"Can I take you somewhere?"

"Over to his job. I reckon they working. They building a house."

"How far?"

"Just a little way."

"Get in, Ma'am." I opened the door to my car.

The lady and the grandkids climbed in and she began to direct me. The youngest, a four-year-old she called Bubba, began to stare at the GPS unit in my car. Bubba leaned over and whispered a couple of times to his older brother.

Bubba turned to me and asked, "Mister, can you turn to Channel 33? I wanna see the cat cartoon."

"Channel 33? No, I can't, Bubba. Can't get it. It's a GPS."

We reached the work site and the lady pointed, "Pull over at that door there and honk the horn. Reckon they working inside today."

I did and a man holding a newspaper over his head walked to my car. I rolled down the window and the lady yelled, "Tell Albert I got car problems."

The man yelled back through the open door, "Y'all tell Albert to get out here!"

This brought out a man, obviously Albert, who listened while the lady explained her problems. He greeted me with, "Damn car."

After much discussion, they decided I'd take the whole crowd, including Albert, back to McDonalds. Absent of other catastrophes, I'd resume my journey to Whiting Field.

Bubba couldn't take his eyes off the GPS. "You sure you can't you turn to Channel 33, mister?"

"Sorry, Bubba. Can't get it."

Albert kept murmuring, "Damn car."

Late that afternoon, after a full day at Whiting, I left to return home. The weather had cleared and nighttime was approaching. I drove through Crestview and chuckled over the morning activity. Curiosity took over and I decided to see what was happening at McDonalds. The old car was gone. I assumed Albert had fixed it. Most likely Bubba was home watching the cat on Channel 33.

Little Pilots

The day was 13 January 2011. My wife and I had been working in preparation for the following Thursday, a week away, specifically to piece together a proper uniform. We had been invited to the Flight Suit Ceremony at NAS Pensacola to celebrate the 100th anniversary of the birth of Naval Aviation.

Marjorie had always worn her mini-pilot wings for formal and other nice functions while I was in the Corps. With all of her searching, without results, it was obvious she had misplaced them. But she was bent on wearing a set for this 20 January ceremony.

Mid-morning she left to drive a cousin to the doctor in Tallahassee. I stayed home to continue getting ready and to make some calls to a few military surplus stores in the Pensacola area to see if they had any Navy mini-pilot wings. The answers kept coming, "No," "Sold out," "I don't know what they are", etc.

On one call, this very sweet-voiced young girl, seemingly trying to be helpful, said, "I'm sorry sir, I don't know what they are. Can you describe them?"

"They are called mini-pilot wings, one and an eighth inches wide. I'm sure you've seen pilots wearing regular sized wings."

"Yes, sir I have, but how can I tell the difference in the ones you want?"

Especially clear I said, "They are called mini-pilot wings. Mini, like small. The regular ones are about two and a half inches wide. Minis are only a little over an inch wide."

After a few long seconds, she came back with, "Well sir, is this something they give to *little* pilots?"

I broke up. I thanked her, said goodbye, and hung up.

Most adversities eventually end up as they should. Later in the day I went to the local hospital for a meeting. My cell phone rang and seeing it was Marjorie, I answered in a whisper. She said, "Hey, I found them! In a little gray box."

So now we were ready and almost in uniform. One day I'll go by that surplus store to thank the young lady.

Bits of Aviation Humor

There are old pilots and there are bold pilots. But there are no old, bold pilots.

If an engine fails on a twin-engine plane, fear not. You will always have enough power to get to the scene of the crash.

We can train monkeys to fly, but they can't make the voice reports.

When a flight is going well, standby; something is about to happen.

The three finest things in life are a good landing, a good orgasm, and a good bowel movement. A night carrier landing is one of the few opportunities in life where, when you do it correctly, you get to experience all three at the same time.

You know that your landing gear is up and locked when it takes full power to taxi to the terminal.

How is an air traffic controller similar to a pilot? Answer: If the air controller screws up, the pilot dies. Likewise, if the pilot screws up, the pilot dies.

A lot of progress is being made in airline flying. Now a flight attendant can get a pilot pregnant.

Airspeed, altitude, and brains are major components of good flying. At least two are required to successfully complete a flight.

What is luck? A smooth landing is luck. Two smooth landings in a row is luck. Three in a row is a damn lie.

Never fly in the cockpit with someone braver than you.

If something hasn't broken on your helicopter yet, it's about to.

Basic Flying Rules: (1) Stay in the middle of air. (2) Do not go near the edges of air. The edges of air can be recognized by the appearance of ground surface, buildings, seas, trees and interstellar space. It is difficult to fly there.

EPILOG

God bless every young man and woman who was ever willing to put his or her life on the line when given the opportunity to protect this country from the enemies of right.

-e-

CPSIA information can be obtained
at www.ICGtesting.com
Printed in the USA
LVOW12*0101071017
551544LV00005B/51/P